Palgrave Studies in Climate Resilient Societies

Series Editor
Robert C. Brears, Avonhead, Canterbury, New Zealand

The Palgrave Studies in Climate Resilient Societies series provides readers with an understanding of what the terms **resilience and climate resilient** societies mean; the best practices and lessons learnt from various governments, in both non-OECD and OECD countries, implementing climate resilience policies (in other words what is 'desirable' or 'undesirable' when building climate resilient societies); an understanding of what a resilient society potentially looks like; knowledge of when resilience building requires slow transitions or rapid transformations; and knowledge on how governments can create coherent, forward-looking and flexible policy innovations to build climate resilient societies that: support the conservation of ecosystems; promote the sustainable use of natural resources; encourage sustainable practices and management systems; develop resilient and inclusive communities; ensure economic growth; and protect health and livelihoods from climatic extremes.

More information about this series at
https://link.springer.com/bookseries/15853

Pedro Henrique Campello Torres ·
Pedro Roberto Jacobi
Editors

Towards a just climate change resilience

Developing resilient, anticipatory and inclusive community response

Editors
Pedro Henrique Campello Torres
Institute of Energy and Environment
Universidade de São Paulo
São Paulo, Brazil

Pedro Roberto Jacobi
Universidade de São Paulo
São Paulo, Brazil

ISSN 2523-8124   ISSN 2523-8132   (electronic)
Palgrave Studies in Climate Resilient Societies
ISBN 978-3-030-81621-6   ISBN 978-3-030-81622-3   (eBook)
https://doi.org/10.1007/978-3-030-81622-3

This Palgrave Macmillan imprint is published by the registered company Springer Nature Switzerland AG
The registered company address is: Gewerbestrasse 11, 6330 Cham, Switzerland

There is no definitive defeat or triumph.

—*Pepe Mujica*

Foreword

The low ambitions of global climate governance, coupled with the conservative pushback against effective climate mitigation, have significantly shaped climate justice debates. Simplistic narratives, which compare the "costs" of climate action to its incalculable benefits, have stalled genuine progress on climate mitigation. As a result, climate activists have reasons to celebrate even conservative policy actions, such as the net-zero targets announced by various world leaders. At the same time, the beneficiaries of the status quo seek to garner public support for unproven offset programmes, often by using progressive sounding labels like "nature-based solutions," which are then embellished by promises of respecting the rights of indigenous and other rural communities. Yet, none of these solutions guarantee climate justice. On the contrary, in most likely scenario, the dominant approaches to climate action, including carbon offsets, net-zero, nature-based solutions and carbon dioxide removal, are more likely to aggravate climate injustices.

The programmes and policies meant to offset carbon emissions linked to luxury emissions of the super-rich lead to the distortion of priorities for the use and management of forests, pastures, mangroves and other natural resources that are crucial for the lives and livelihoods of the poor and the marginalized majorities. As such, climate solutions that powerful political and economic actors in the Global North favour often perpetuate new forms of economic imperialism and neocolonial control of the global periphery. We have witnessed the emergence of "carbon colonialism," in which the interests and ideas of powerful actors in the Global North determine the regulation and management of carbon emissions and stocks, leading to the subjugation of the interests and aspirations of people in the Global South. Market actors in the Global North perpetuate such exploitative systems of carbon control in active collaboration with the political elite in the Global South. The goals and strategies of climate justice must include confronting these transnational elite networks, which facilitate the coercive takeover of resource management authority. Ironically, such a takeover produces perverse incentives that encourage extractive resource use that is unlikely to be sustainable or socially just. The United Nations Declaration on the Rights of Peasants and Other People Working in Rural Areas (UNDROP) prohibits such coercive and counterproductive take over resources, including in situations where powerful actors seek to justify these actions on the pretext of biodiversity conservation or climate mitigation. Similar global conventions are needed to prohibit land and resource grabs in urban and rapidly urbanizing regions.

A more vicious form of racially defined domination occurred via climate denialism, leading to unacceptable delays in the long-overdue mitigation actions. Powerful transnational actors, e.g. the fossil fuel corporation, orchestrated a well-funded, multi-decadal campaign to dismiss scientific evidence on anthropogenic climate change and sow seeds of doubt among the public. Those responsible for climate denialism and sabotaging progress on climate action must be held accountable for their misdeeds. Considering the enormity of the costs that global climate crisis imposes on the countries of Global South, especially the poorest people within those countries, climate denialists should face trials for crimes against humanity. Holding the powerful actors responsible and accountable will send a powerful message for others who are still in two minds about what the climate-changed world means for their business models. Such global actions—spurring aggressive climate mitigation and holding the powerful actors accountable for their reckless behaviour—are necessary to pursue climate justice. Yet, even such remarkably ambitious global actions will not be *sufficient* for the realization of climate justice. The editors and contributors of this volume remind us that climate justice requires engaging with an entire suite of problems and solutions in sectors as varied as forests, fisheries, drinking water supply, climate action plans and intersectional vulnerabilities that result when the climate crisis intervenes in politically unstable regions that have also been at the receiving end of transnational extractive regimes. For example, access to safe drinking water in the coastal regions of Bangladesh fighting salinity ingress requires setting up community-based desalinization systems, rain-water harvesting structures and intervening in the management of irrigation systems. Protecting the precarious livelihoods of subsistence fisher folks in Uruguay requires adapting the fish gear to prevent sea lions from feeding on the fish caught in fishlines and damaging the equipment in the process. It also needed creating broader public awareness about the social, cultural and ecological contributions that small-scale subsistence fishing makes. In Brazil, which is home to a well-organized civil society and some of the most prominent social movements of the poor, city administrators and policymakers have failed to promote broad-based social engagements around the development of climate action plans. Climate justice in this case requires confronting these exclusions, making the policymaking process more inclusive and developing specific elements of climate action that the previously constituencies may prioritize. On the other hand, climate justice for indigenous communities in the Solomon Islands requires controlling the destruction caused by extractive industries of logging and mining while protesting indigenous communities against the risks involved in forest conservation projects funded by carbon offsets projects, which are subject to the vagaries of speculative transnational capital flows.

The case studies and analyses presented in this volume point to the indispensable role of programme design and the development of physical infrastructure for progress towards climate justice. More research and scholarship are needed to explore the multiple ways in which international and transnational demands for climate justice shape the local realization of these ideals. In the current global system founded on the principles of national sovereignty, engaging with national and subnational governments is necessary for fostering climate justice interventions in diverse sectors. The success of these engagements depends on the types of political intermediation mechanisms (PIM) prevalent in different countries. As I elaborate in *Democracy in the Woods*, PIMs are well-established processes and relationships that help citizen groups, civil society organizations and social movements engage in political and policy processes that affect them directly. In many instances, the nature of PIMs may matter more than formal institutions of democracy, which are vulnerable to majoritarian politics. Effective PIMs can act as conduits of transnational solidarity actions and may even induce the so-called boomerang effect of pressurizing national and subnational governments to act appropriately. We have strong reasons to believe that an exclusive focus on international climate justice, in a system dominated by national elites, may have unwittingly undermined the goals of bringing justice to the world's people.

This volume highlights the importance of multi-sectoral, multi-scale, transnational engagement with climate impacts and climate action. Such eclectic engagements are essential for developing interventions for translating the moral and ethical arguments in favour of climate justice into concrete and immediate gains for the world's poor and marginalized. At its core, climate justice is a political question, and engaging with the politics of the place will be crucial for the long-term success of any efforts on this front.

Let thousand climate justice interventions bloom in the hills, valleys, deserts, villages, suburbs and cities where we live, work and play.

Storrs, CT, USA   Prakash Kashwan
June 2021

Acknowledgements

This book was written during the most significant health crisis in our history: the COVID-19 pandemic. We dedicate this work to the victims of the Sars-Cov-2 virus, especially the most vulnerable populations that have not only been more affected but have less capacity to respond to the crisis. In the same way, we are extremely grateful to the authors who, despite the challenges imposed by COVID-19, delivered their contributions.

This book is also dedicated to all those who believe that South–North, North–South learning is possible. Decolonizing the imaginary, new governance and planning practices are essential in searching for a just and sustainable present and future. We also thank all traditional communities, first nations, urban and rural community organizations for keeping alive the flame that another *transition* is possible.

This work was also carried out with the aid of a grant from the The São Paulo Research Foundation (FAPESP) grant number 2018/06685-9. And it is part of the project *COPPLANNING—Community-Based Planning and Participation for Low Carbon Transition: A Global South-North Comparative Study Centring Traditional and First Nations Communities*

(FAPESP grant. 2019/23559-0), hosted by University of Queensland (Australia) and University of São Paulo (Brazil), led by Professor Silvia Helena Zanirato and Professor Kristen Lyons.

Contents

List of Figures

List of Tables

1

Introduction: Community Practices and Climate Justice—A Global South Perspective in the Face of Climate Emergency

Pedro Henrique Campello Torres
and Pedro Roberto Jacobi

Abstract How to address climate justice and community responses to climate emergencies? How to contribute to filling knowledge gaps in the global south concerning data on informal settlements and adaptation action for vulnerable people who live in this growing territory? The Global South, the focus of the present book, has an agenda of environmental struggles and priorities that need to be integrated into a common agenda against environmental inequalities and environmental justice. Sharing knowledge and practices are an essential path towards a just sustainable planet. That's the main book objective, focused on sharing practices/proposals or methodological insights for working on participatory/community planning in regions of the Global South in the face of climate change.

P. H. Campello Torres (✉) · P. R. Jacobi
Institute of Energy and Environment, University of São Paulo, São Paulo, Brazil
e-mail: pedrotorres@usp.br

© The Author(s), under exclusive license to Springer Nature Switzerland AG 2021
P. H. Campello Torres and P. R. Jacobi (eds.), *Towards a just climate change resilience*, Palgrave Studies in Climate Resilient Societies, https://doi.org/10.1007/978-3-030-81622-3_1

Keywords Climate justice · Global south · Knowledge gaps · Share knowledge

Background

This book seeks to contribute to studies and researches on climate justice and community responses to climate emergency. Bai et al. (2018) suggested six research priorities for cities and climate change. The authors state that the most significant knowledge gaps would be in the Global South, mainly concerning data on informal settlements"sparse or non-existent." Besides, the authors draw attention to the fact that by 2050, three billion people, mainly in the Global South, will be living in slums, dramatically exposed to the effects related to climate hazards such as floods—and water scarcity, I would add. This publication also dialogues with the author's understanding on "Enabling these communities to adapt is a priority," as well as we must tackle poverty and adaptation at once (Pelling & Garschagen, 2019).

This book also answers the call of Nagendra et al. (2018) who consider that challenges, and opportunities—related to urban-environmental challenges—are found in the Global South. But as the authors say "current urban knowledge is predominantly shaped by research on and from the global north." Chapters gathered in this book also address authors calling for a renewed research focus on urbanization in the south and suggest targeted efforts to correct structural biases in the mainstream knowledge production system.

In this sense, this book's authors work directly on concrete cases that elucidate crucial themes related to resilience, adaptation and the imperative active participation of communities. The novelty in this book is that all the cases covered in it are from the Global South's territories. Not only a country or a region. But it is covering all the continents where territories and their populations are located in the Global South: Brazil and Uruguay (Latin America), Pakistan (Asia), Mozambique (Africa), Solomon Islands (Oceania).

Despite that, it is worth noting that the Global South is not homogeneous, as some studies indicate when generalizing this region. The objective of bringing cases from the different areas of this politically produced territory precisely illuminates this issue. The Global South, which is often used mechanically and homogeneously, is highly heterogeneous in all its spheres and layers. For example, Brazil's colonial historical formation is not the same as in Pakistan, Mozambique, other territories of the African continent, or Salomon Island, in the Pacific.

It is recommended to pay attention to the territories' specificities and decolonize our understanding of the regions. In Brazil, permanences and continuities of past colonial domination—structural racism, slavery, patriarchalism—and other material and symbolic forms of power by local elites contributed to territories unequal social production. This long-lasting historical process perpetuates the chains of inequalities that will need to be destroyed if new paths are to be followed.

The Global South, the focus of the book, has an agenda of environmental struggles and priorities that need to be integrated into a common agenda against environmental inequalities and environmental justice. But the importance of this agenda will be overwhelming, not if it is a mere mirror developed in the North, but if it is built from below, as an outcome of its specificities and local demands. It may seem obvious, but it is not the current practice. On the contrary, it tends to be neutralized with emerging agendas coming from the North through NGOs, multilateral institutions, corporations and academia. It is also part of a neoliberal project that seeks to neutralize understandings and imaginaries: homogenizing interpretations and paths of fighting against elitist, monopolistic and individualist practices.

That is why a transition project to a low carbon economy cannot be carried out with support for coups d'état as in Bolivia in 2019, due to the extractive interest in lithium. Or with severe social impacts with the artisanal exploration of Cobal—an essential element within supply chains are driving the so-called energy transition to electric vehicles in Europe-US—in mines in the Democratic Republic of Congo (Nkumba-Umpula et al. 2021). Nor should they serve the interests of large corporations and well-known billionaires in the quest to renew their productive capacity and profits. A Green New Deal or a "Green Recovery," as it has been propagated on all continents, cannot be used only to renew the system itself, under the discourse that it is green. It needs to seek to tackle the overcoming and reduction of poverty and inequality in the first place.

That is why illuminating insurgent, non-state, community practices, traditional knowledge is fundamental and the main objective of
this book. As Torres et al. (2020) showed, the climate justice agenda
continues to spread slowly in the Global South. The cartographic representation (Fig. 1.1) of the places where the most 2019 Climate Strikes
protesters were concentrated shreds of evidence in an empirical way.
It does not mean that climate justice practices based at a community
level are not taking place. Here, it is essential to note a statement by
Kashwan (2021) for whom it is necessary to differentiate the "Climate
Action Movements" that should not be understood to be synonymous
with "Climate Justice Movements." In the specificities of the heterogeneous Global South, this difference is relatively straightforward, and
the chapters that have been assembled and that invoke an approach for
environmental justice reinforce this understanding. One thing is climate
injustice: the unequal impacts of climate change and the unequal ways
and capacities to react and adapt to them. Another is local, national, or
global actions and movements with the climate agenda, led, for the most
part, by international non-governmental organizations or foundations.
There is yet another dimension still being built in the Global South.
From below, seeking to rotate and deconstruct a colonial logic and past,
instigating demands for historical rights and reparations—aligned with

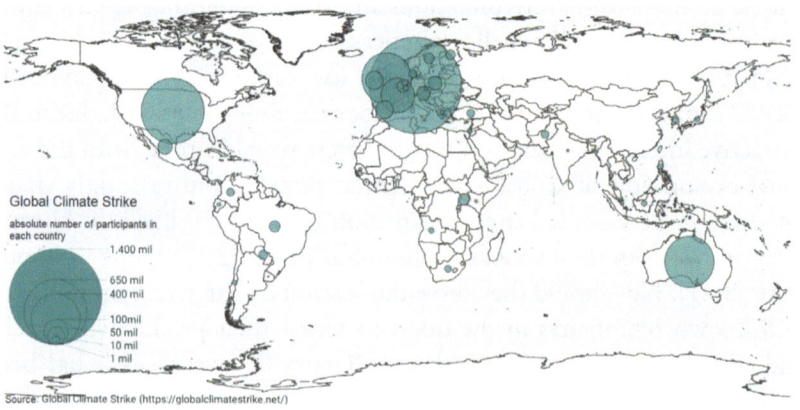

Fig. 1.1 Population participation in the climate strike by country, in absolute number of participants (*Source* Torres et al. [2020])

the "Climate Justice Movements" and emerging themes such as climate apartheid, climate immigration, climate gentrification, and many other inequalities by which climate justice movements must truly target their actions on their scale and possibilities.

But then, what are climate justice? Climate injustices are the disproportionate exposures that most vulnerable groups or communities suffer in a given territory. The central idea here is that everyone will feel the effects and impacts of climate change in the four corners of the planet, but they will not feel these effects in the same way. Nor will they react to these effects in the same way. These asymmetries we call environmental inequalities, climate inequalities, and environmental privileges. As a counterpoint to these injustices, we call Climate Justice. Regarding the injustices produced in countries in the Global South, such as the ones presented in the book, the tendency is to be even more severe than in countries in the Global North. The colonial past, slavery, structural racism, patriarchalism, and an agrarian-exporting landowner elite did not collapse with independence, the proclamation of the republic, or restoration of democracy after the military dictatorship. On the contrary, they bring marks of historical continuities even in processes of discontinuities. Extreme weather events, such as floods, typhoons and tornadoes, prolonged droughts, and heatwaves, have appeared more frequently and intensively in all parts of the planet. Impacts have the potential to reach everyone, regardless of their region, nationality, ethnic group, or gender. However, not everyone suffers or will suffer these impacts in the same way. And, more than that, the forms and capacities of planning, reacting, and adapting are not or will not be the same for everyone. These unequal and disproportionate forms of impacts are called Climate Injustices. The reaction to these injustices we call Climate Justice.

The first international forum for Climate Justice took place in the Netherlands, in 2000, in the city of The Hague, in parallel with the UN Climate Conference (UNFCCC—COP6). Two years later, in Johannesburg, South Africa, together with the Rio+10/Earth Summit, the Bali Climate Justice Principles were adopted. The international movement for Climate Justice has grown exponentially in these years, gaining strength, especially in meetings parallel to official events of the United Nations or other forums. It was like that in 2004 with the founding of the Durban

Group for Climate Justice, the founding of the global Climate Justice Now! Coalition, as well as the Global Humanitarian Forum, the Climate Justice Action in 2009. The year 2009 is symbolic for global movements for Climate Justice as it marked the exclusion of social movements and non-governmental organizations from the official UN summit, held in Copenhagen, Denmark. The solid military repression and the considered failure of the diplomatic climate agreement have intensified the separation between social groups and their demands. In response, the World Conference of Peoples on Climate Change and the Rights of Mother Earth were organized in 2010 in Tiquipaya, 30 kilometres from Cochabamba, called the counter-summit (in reference to the official UN summit). Seeking to break with formal diplomacy, one of Cochabamba's main principles was the understanding that it would not be possible to face climate change without questioning the capitalist system on a planetary scale. Hence, the slogan "Change the system, not the climate."

But this is not a book of denunciations of climate injustice. In addition to exposing the impacts and inequalities, it seeks to expose examples of participatory/community planning in regions of the Global South in the face of climate change. It is, therefore, a living research object— in movement. And sharing and exchanging knowledge, limits, barriers, challenges, and successes are essential in the face of the challenges of our times.

Scope of the Book

To achieve the objective described above in contributing to studies and researches on climate justice and community responses to climate emergency in the Global South, a group of specialists from different parts of the planet was brought together. Gender balance, regions, experts, experience with concrete case studies on the field were taken into account to cover the expertise needed. The result is a truly interdisciplinary and multi-ethnic effort of dialogue and learning.

Each case is an invitation to the reader to know a little more about the local reality of each object studied, anchored, on the other hand, in global issues. Fishers, First Nations, Traditional Populations are some examples of the plunge into the universe addressed by the authors, from Uruguay to Pakistan, from the Solomon Islands to Brazil, with a stopover in Mozambique.

The authors' call sought contributions that focused on sharing practices/proposals or methodological insights for working on participatory/community planning in regions of the Global South in the face of climate change. Research on building a more resilient response to climate change adaptation, focus on inequality, vulnerability and a disaster justice perspective were the primary intended goals.

Structure and Contents of the Book

This book is divided into five chapters developing cases, one for each Global South region, for a more resilient and communities responses to climate change. In addition, the last chapters provide a synthesis and ways forward to strengthen community practices and climate justice from the Global South.

Chapter 2 **Building Small-Scale Fishers' Adaptive Capacity Through Participatory Action Research in Coastal Uruguay**, from Micaela Trimble, Paula Santos addresses the issue of small-scale fisheries with a case study of Piriápolis, coastal Uruguay, where a multistakeholder participatory action research (PAR) group was formed in 2011 to address local problems of the fishery in a transdisciplinary manner. In this chapter, the authors analyse the contributions of this PAR to building five domains of fishers' adaptive capacity: assets, flexibility, social organization, learning and agency (as proposed by Cinner et al., 2018). The study provides evidence supporting the potential of participatory approaches for building adaptive capacity and climate resilience. The findings also show that tPAR contributed to the development of adaptive capacity in an integral approach since the five domains were positively impacted.

Ensuring Water Security for Climate Vulnerable Communities in Coastal Bangladesh, Chapter 3, from Istiakh Ahmed, Saleemul Huq, Shahrin Mannan, Juel Mahmud, Mizan R Khan expose the case of two coastal villages in Bangladesh, where climate change is likely to increase river salinity leading to shortages of drinking water and irrigation system at the coastal belt. Approximately 20 million people live along the coastline and frequently rely on pond water which are primarily rain-fed, but also mix with river water, soil run-off and shallow groundwater. The study intends to identify specific vulnerabilities experienced by different groups in these coastal villages, which share similar geographic and socioeconomic characteristics. It examines the complex interlinkage of water security and explores how different environmental stressors and other factors are negatively affecting water security. Based on the findings, it also proposes recommendations to be made both to decision-makers at the national level for development of policies and plans to enhance the resilience of coastal communities.

In Chapter 4, **Policy and Planning Responses to Climate Change in Solomon Islands: A Place for Forest-Based Carbon Offset Initiatives in Building Just and Resilient Territories?** Kristen Lyons and Peter Walters address Pacific Island nations on the front lines of some of the worst climate change impacts. As small states, these nations also frequently take leadership roles in local livelihood, policy and political advocacy responses to climate change mitigation and adaptation. This chapter takes the case of Solomon Islands, an archipelago comprising over 900 islands and former British colony that won independence 1978, to apply an environmental justice approach in examining some of these responses. Drawing lessons from these initiatives, this chapter assesses the place of forest-based carbon offset initiatives in building just and resilient territories in the face of powerful global economic interests. Authors conclude that distributive and procedural justice will be vital if carbon offset initiatives are to be part of a climate just future.

Beyond Climate: How to Approach Resilience in Extremely Vulnerable Territories? Chapter 5, from Carla Gomes and Luísa Schmidt, analyzes the rising of climatic and political instability, in a region that was already affected by persistent poverty despite positive growth trends at the national level. Cyclone Kenneth and the COVID pandemic, amid rising political violence, have afflicted Cabo Delgado populations and aggravated the situation of those most vulnerable, posing a very significant threat to the fulfilment of the Sustainable Development Goals in this region. In that sense the chapter analyses how current adaptation policies and plans in Mozambique respond to the challenges of protecting human capabilities and improving social resilience. The authors undertake this through a review of the literature and a qualitative analysis of policy instruments in the fields of rural development, climate action and poverty reduction. The chapter then offers recommendations for new adaptation policies, including the National Adaptation Plan under the Cancún Agreement, as well as an integrated approach to climatic and post-COVID resilience which supports the fulfilment of the SDG amongst these critically vulnerable communities.

Pedro Henrique Campello Torres, Ana Lia Leonel and Gabriel Pires de Araújo, in Chapter 6, **Climate Injustice in Brazil: What We Are Failing Towards a Just Transition in a Climate Emergency Scenario?** question how climate justice is—if it is—present in the country climate change debate and praxis. For this purpose, they analyse three climate change plans to see how categories related to justice, poverty and right issues are inserted or not inserted in current local planning practices.

Chapter 7 **Community Practices and Climate Justice from the Global South—Synthesis and Ways Forward**, from Pedro Roberto Jacobi and Pedro Henrique Campello Torres, closes the book offering the reader a systematization of the lessons learned from the cases and future paths to be explored by research agendas. Impacts and capabilities to react to extreme events are uneven, and vulnerable groups are more susceptible to climate injustices that affect different regions of the planet. As a counterpoint to climate injustices, social movements defend the struggle for Climate Justice as a way to fight inequalities, denounce major polluters—whether corporations or countries—as well as false narratives and solutions that end up perpetuating privileges instead of fighting them.

References

Bai, X. R., Dawson, R. J., Ürge-Vorsatz, D., Delgado, G. C., Barau, A. S., Dhakal, S., Dodman, D., Leonardsen, L., Masson-Delmotte, V., Roberts,

D. C., & Schultz, S. (2018). Six research priorities for cities and climate change. *Nature, 555*(7694), 23–25. https://doi.org/10.1038/d41586-018-02409-z.

Cinner, J. E., Adger, W. N., Allison, E. H. et al. (2018). Building adaptive capacity to climate change in tropical coastal communities. *Nature Climate Change, 8*, 117–123. https://doi.org/10.1038/s41558-017-0065-x.

Kaswan, P. (2021). Climate justice in the global North: An introduction. Case studies in the environment. https://doi.org/10.1525/cse.2021.1125003.

Nagendra, H., Bai, X., Brondizio, E. S., & Lwasa, S. (2018). The urban south and the predicament of global sustainability. *Nat Sustain, 1*, 341–349. https://doi.org/10.1038/s41893-018-0101-5.

Nkumba-Umpula, E., Buxton, A., Schwartz, B. (2021). *Islands of responsibility? Corporate sourcing of artisanal cobalt in the Democratic Republic of Congo.* IIED, London.

Pelling, M., & Garschagen, M. (2019). Put equity first in climate adaptation. *Nature, 569*, 327–329.

Torres, P. H. C., Jacobi, P. R., & Leonel, A. L. (2020). Nem leigos nem peritos: o semeador e as mudanças climáticas no Brasil. *Revi Política Soc, 44*, 34–52

Torres, P. H. C, Leonel, A. L., de Araújo, G. P., & Jacobi, P. R. (2020, April). Is the Brazilian national climate change adaptation plan addressing inequality? climate and environmental justice in a global south perspective. *Environmental justice, 13*(2), 42–46. https://doi.org/10.1089/env.2019.0043.

2

Building Small-Scale Fishers' Adaptive Capacity Through Participatory Action Research in Coastal Uruguay

Micaela Trimble and Paula Santos

Abstract In Uruguay, as globally, small-scale fisheries sustain numerous communities. Changes in wind patterns and ocean warming in the southwest South Atlantic affect small-scale and large-scale fisheries. This chapter analyses the small-scale fishery of Piriápolis, Uruguay, where a multistakeholder participatory action research (PAR) group was formed in 2011 to address local problems. The objective was to study the contributions of this PAR to building five domains of fishers' adaptive capacity. The findings show that the PAR contributed to integrally developing adaptive capacity since the five domains were positively impacted: national financial support was obtained, there was flexibility in creating a new fishing gear, new commercial skills were developed, relationships

M. Trimble (✉)
South American Institute for Resilience and Sustainability Studies (SARAS), Maldonado, Uruguay

P. Santos
Universidad de la República, Montevideo, Uruguay

© The Author(s), under exclusive license to Springer Nature Switzerland AG 2021
P. H. Campello Torres and P. R. Jacobi (eds.), *Towards a just climate change resilience*, Palgrave Studies in Climate Resilient Societies, https://doi.org/10.1007/978-3-030-81622-3_2

and trust were built throughout the process, participants highlighted learning, and bottom-up institutional changes were triggered.

Keywords Small-scale fisheries · Southwest Atlantic Ocean · Climate change · Resilience · Transdisciplinarity

Introduction

Small-scale fisheries (SSF), also known as artisanal fisheries, sustain numerous communities worldwide (FAO, 2020). They provide a source of income and animal protein for numerous people, and they also have cultural relevance; small-scale fishing is commonly described as "a way of life" (Trimble & Johnson, 2013; Weeratunge et al., 2014). Despite their socioeconomic and cultural importance, SSFs are usually a marginalized sector in many parts of the world, such as in Uruguay.

Climate change is one of the drivers that can intensify the current fisheries crisis of declining catches, among others. Like the agricultural sector, fisheries productivity is likely to be decreased (Cheung et al., 2010; Sumaila et al., 2011). The increase of sea temperature, changes in sea currents and the sea-level rise are among the factors affecting fisheries activities. In the Southwest Atlantic Ocean, one of the most productive marine areas of the world, a recent study showed that in response to changes in near-surface wind patterns, the Brazil current has been intensifying and shifting southwards, leading to ocean warming along its path and the Río de la Plata (Franco et al., 2020). Moreover, there is evidence of impacts of long-term ocean warming on large-scale (or industrial) fisheries in Uruguay (Gianelli et al., 2019b). Small-scale fishers of different coastal locations in this country have also been perceiving changes in climate conditions, such as more frequent wind storms or unfavourable onshore wind conditions, which have led to a reduction in the number of fishable days (Gianelli et al., 2019a; Trimble, 2013).

According to the Intergovernmental Panel on Climate Change (IPCC) (2014, p. 128), vulnerability is "the propensity or predisposition to be adversely affected. Vulnerability encompasses a variety of concepts and elements including sensitivity or susceptibility to harm and lack of capacity to cope and adapt". Most vulnerability assessments consist of three main components: exposure (local manifestations of climate change), sensitivity (the degree to which people depend on affected resources) and adaptive capacity (ability that allows humans, as individuals or as a group, to respond to changes and take advantage) (Adger, 2006). The two former dimensions are positively correlated to vulnerability, whereas measures that promote adaptive capacity can reduce vulnerability.

Adaptive capacity is usually measured following the 5-capital-approach based on the Sustainable Livelihoods Framework, which distinguishes five kinds of assets or capitals in a system: human, social, physical, financial and natural (e.g. Pandey et al., 2017; Williges et al., 2017). However, recent studies have shown that adaptive capacity is also about the capability to convert these resources into action (e.g. Cinner et al., 2018; Coulthard, 2012). A synthesis research by Cinner et al. (2018, p. 117) across a range of disciplines proposed that adaptive capacity consists of five key domains: "(1) the assets that people can draw upon in times of need; (2) the flexibility to change strategies; (3) the ability to organize and act collectively; (4) learning to recognize and respond to change; and (5) the agency to determine whether to change or not".

In this chapter, we analyse a case study in Piriápolis, coastal Río de la Plata—Uruguay, where a multistakeholder group was formed in 2011 through participatory action research (PAR), involving small-scale fishers, scientists, government and non-government actors, with the purpose of addressing local problems of the fishery in a transdisciplinary manner, while also contributing to laying the basis for small-scale fisheries co-management in the area (Trimble & Berkes, 2013). The goal is to analyse the contributions of this PAR process to building fishers' adaptive capacity, following the five domains proposed by Cinner et al. (2018).

A Fisheries Case Study in Piriápolis, Coastal Uruguay

Piriápolis is a coastal tourist city located in Maldonado Department, 98 km East from Montevideo (the capital of Uruguay). Small-scale fishing is an important activity throughout the year. Fishers use gillnets and long-lines onboard their boats (4–8 m long) to catch a variety of fish species (such as whitemouth croaker and Brazilian codling), which they sell to middlemen (mostly), restaurants, or directly to the public. The number of small-scale fishers has decreased over the years since the activity is not as economically profitable as in the past; catches have been declining, fishing zones are farther, etc.

Fishers who remain in the activity have managed to cope with these challenges, by diversifying their fish sales methods (e.g. not depending entirely on middlemen, who set low prices), and sometimes diversifying also their sources of income (e.g. not depending only on the fisheries sector).

In 2011, in the context of the first author's doctoral research, a PAR process was initiated by inviting multiple stakeholders to contribute collectively to the solution of problems of the local fishery: small-scale fishers from Piriápolis, members of the National Directorate for Aquatic Resources (DINARA, within the Ministry of Cattle, Agriculture and Fisheries, in charge of fisheries management), academia and civil society (Trimble & Lázaro, 2014). These actors formed the so-called POPA Group ("Por la Pesca Artesanal en Piriápolis"—For Artisanal Fisheries in Piriápolis).

Three main problems were addressed by this multi-stakeholder group. First, sea lions feed from fish entangled in gillnets and long-lines, reducing fishers' catches and causing damage to the fishing gear. Second, around 2010, the increasing sale of pangasius (also known as basa) started to affect the sale of fish caught by the small-scale fishing sector. Uruguay began to import this catfish in 2008 from aquaculture producers in Vietnam at very low cost. Pangasius sales in Piriápolis (and other parts of the country) increased from that moment onwards in local markets, supermarkets and restaurants, where the imported fish was often sold as if it were a species caught locally, deceiving the consumer. Third, the broader social problem of the low valorisation of the fishers. Small-scale fishers stand out for their knowledge of the sea and their practices which, transmitted from generation to generation, contribute to environmental sustainability and food security, among others (McGoodwin, 2001). However, their potential societal contributions are neglected and underappreciated (Song, 2018). Fishers in Uruguay (and the world) who experience situations of vulnerability, have expressed the need to become valued partners in fisheries management (Trimble & Johnson, 2013).

Among the accomplishments of the group, over six months of hard work, POPA organized the First Small-Scale Fisheries Festival (or Show). It took place during a summer weekend in 2012 in Piriápolis, as a public communication activity which reached around 3,000 people, promoting the valorization of small-scale fisheries and the consumption of local fish species (Fig. 2.1).

Interviews with fishers and non-fisher members of the group were conducted as part of different graduate theses, to assess the PAR process and outcomes, the contributions to fisheries co-management and the implications for fishers' social well-being (e.g. Santos, 2017; Santos et al., 2021; Trimble, 2013; Trimble et al., 2014). A more recent study proposed indicators to assess the vulnerability (exposure, sensitivity and adaptive capacity) of small-scale fisheries to climate change in coastal Uruguay, taking Piriápolis as a case study and conducting interviews with 18 fishers (Teruggi, 2019). The indicators of adaptive capacity included the diversification of sales methods, the diversification of target species, the reliance on the sector, the use of meteorological forecast, the port infrastructure, the access to credit and insurance and social integration (referring to fishers' involvement in associations, social groups, unions, etc.).

Building Fishers' Adaptive Capacity Through Participatory Action Research (PAR)

Based on the findings from our previous studies in Piriápolis, and some references to Teruggi's work (2019), this section analyses how the PAR process contributed to the five key domains of adaptive capacity to climate change: assets, flexibility, social organization, learning and agency (Cinner et al., 2018). A summary of this analysis can be found in Table 2.1.

Fig. 2.1 **a** Fishers were interviewed by the media during the First Small-Scale Fisheries Festival in Piriápolis. **b** Fisherwoman explaining the use of long-lines to the public of the Festival

Table 2.1 Contribution of the PAR process in Piriápolis for building fishers' adaptive capacity (based on the five domains proposed by Cinner et al., 2018)

Adaptive capacity domains (Cinner et al., 2018)	Changes triggered by the PAR process (Santos & Trimble, 2018; Santos et al., 2021; Trimble & Berkes, 2013; Trimble & Lázaro, 2014)
Assets	
People may have access to financial, technological and service (health care, among others) resources, which are considered as assets	National financial support (>US$100.000)
Flexibility	
Flexibility from switching between adaptation strategies favours adaptive capacity to climatic impacts	Innovation in developing an alternative fishing gear for the area (fish traps) Development of new commercial skills
Social organization	Formation of a multi-stakeholder group (POPA)
Cooperation and collective action can be favoured (or inhibited) by the ways in which society is organized	Improved relationships Trust building among stakeholders
Learning	Learning to listen to others
Multiple organizational, spatial and temporal scales enable experimental or experiential learning	Relating to others Public participation Priority to group interests over individual ones
Agency	Catalysed social capital
Free choice to respond to environmental change	Renewed social representation Triggered bottom-up institutional change

Assets

Financial, technological and service (such as health care) resources are considered as assets that people have access to, individually or collectively (Cinner et al., 2018). These shall position people to better adapt in conditions of change.

In Piriápolis, as in most coastal Uruguay, the weak presence of formal organizations, public or private, which can grant loans or microcredit to fishers, have generated an informal debt market in which middlemen lend fuel, nets, baits or money to fishers. This activity, developed especially in fish scarcity periods, establishes a relationship of strict dependency between fishers and middlemen, who want to have the guarantee of future catches to sell to the processing industry. Consequently, in these situations, fishers are not free to efficiently allocate their catch: even if it would be more profitable for the formers to sell their product straight to consumers or restaurants, they can only trade with middlemen. During the fishing season, when the catches are voluminous, the middleman is normally the first choice for 72% of fishers (Teruggi, 2019).

However, the PAR offered a different form of financial support through project funding directly received from the State to address, through participatory research, problems identified by the fishers. With the successful precedent of the Fisheries Festival in 2012, DINARA (through an agreement with the National Agency for Research and Innovation of Uruguay) funded a research project (50,000 USD, 2013–2014) submitted by POPA to investigate how to mitigate the interaction between small-scale fisheries and sea lions, the first problem identified. Fish traps specifically tailored according to the characteristics of the local fishery were tested as alternative fishing gear.

After this, the General Directorate for Rural Development of the Ministry of Cattle, Agriculture and Fisheries of Uruguay funded another POPA project (50,000 USD, in 2015–2016) to advance results of the previous one. Through this project, an adaptation of the fish traps was locally developed and tested in a participatory manner (Santos et al., 2021).

Flexibility

According to Cinner et al. (2018), switching between adaptation strategies provides individuals with more flexibility to better adapt to climatic impacts. During a five-year period, the PAR process challenged the capacity of fishers to develop a variety of tasks and undertake varied initiatives to contribute to the solution of the three problems identified.

To address the problem of the interaction between the fishery and sea lions, with the public funding received, the fishers developed a new fishing technology to minimize the impact (i.e. the fish traps). The design was done coordinately, involving fishers, a technical team of DINARA and the researchers of POPA. Fishers were able to create, execute, adopt and monitor the performance of the locally designed fishing traps. An additional task was to share this technology with other fishing communities of the Río de la Plata coast.

The PAR process required the fishers to develop new strategies to promote their catch for direct sale to restaurants of Piriápolis. As a result, in 2016, the launching to the press of small-scale fisheries value materials took place in two local restaurants simultaneously, organized by POPA. These actions resulted in strengthened commercial relations between the fishers and the restaurant owners.

Social Organization

The ways in which society is organized to enable (or inhibit) coop-eration, collective action and knowledge, as defined by Cinner et al. (2018), will determine adaptation to climate change. Indeed, the PAR fostered formal and informal relationships between its members, with the community, and other organizations (including the State, the private sector and the media), joining efforts with the fishers to deal with prob-lems they identified, and providing them with social support and access to knowledge and resources. More particularly, the PAR process knitted relationships based on trust and respect among its members (Trimble & Berkes, 2013). Trust is an essential part of the social capital that needs to develop among a group of people trying to solve a problem and trust lubricates collaboration (Pretty & Ward, 2001).

Research undertaken by Trimble and Berkes (2013) showed that during the first two years of the PAR process most relationships between and within the participating stakeholder groups improved (including relationships formed during the process), and none became worse. In addition, trust and respect among participants increased in most relation-ships. In particular, participatory research contributed to building condi-tions for co-management by bringing together fishers and DINARA, facilitating their dialogue and enabling a more direct relationship.

From a sample of Piriapolis' fishers further interviewed by Teruggi (2019), as an indicator of adaptive capacity to climate change (namely social integration), only half of them responded that they take part in associations. Fifty per cent mentioned participating in the POPA Group (although the interviews took place in 2018, after one year of group inactivity). The rest split between Karumbé—a national NGO devoted to sea turtle and marine conservation (20%), SUNTMA—the national union of sea workers (20%), and the Local Fisheries Council for Co-management (10%)—established in 2015.

Learning

The PAR process offered a platform for developing fishers' capacity to generate and process new information about fisheries problems and their possible solutions. According to Trimble and Lázaro (2014), participatory research can be seen as a way of fostering learning and co-production of knowledge, both tending to deal with uncertainty, augmenting the resilience of social-ecological systems. In addition to the new knowledge which contributes to understanding or solving a local problem, learning among participants is another key outcome of participatory research.

The learning process in the Piriápolis case was experimental and experiential and occurred within and across multiple organizational, spatial and temporal scales. All participants stated that they learned throughout the participatory research process, namely to listen to others, to relate to different people, to work through a participatory process, to discuss and debate in an organized way and to prioritize the group interest over the individual, among others. Other types of learning included specific information about the discussed topics (Trimble & Lázaro, 2014).

Agency

The power of freedom or the individual, or collective ability of people to have free choice in responding to environmental change, depict the agency component of adaptive capacity (Cinner et al., 2018). The PAR process of the POPA Group can be taken as a clear example of it, as it involves the empowerment of fishers through a participatory process, one of the three key types of actions identified as necessary to build agency for adaptive capacity, along with incorporating local or customary knowledge skills and removing barriers that may be inhibiting people's ability to exercise agency (Cinner et al., 2018).

Fishers who were eager to be part of the PAR initiative freely decided on the three specific problems the group acted upon collectively. During a process of monthly workshops, the team discussed strategies to address these linked social-environmental problems. As a result, the group opted to tackle the second and third problems through the organization of a public communication activity, the First Small-Scale Fisheries Festival. Within the context of the PAR, fishers co-worked or related personally with the Mayor of Piriápolis, the Mayor of Maldonado Municipality, the Director of DINARA and the military, among others, in the organization of this event. The local and national media supported the initiative.

Ten months after the Festival, DINARA released a press communique on the quality of pangasius imported to Uruguay, stating that all imported species to the country systematically must undergo necessary sanitary testing, also alerting consumers to the "fraud" of some fish retailers and restaurants incorrectly labelling the species. The POPA Group had been reporting this situation since the creation of the group to the Head of the Small-Scale fisheries Unit of DINARA, as he was a member of POPA then, and at the time of this first official announcement (Santos et al., 2021).

PAR in the Science-Policy-Society Interface in the Context of Global Climate Change

Climate change issues involve multiple complexities and uncertainties. Therefore, for addressing these and for bridging the adaptation gap, it is recommended to involve the different stakeholders or actor groups. One approach for doing so, in the science-policy-society interface, is PAR or participatory research, in which resource users or community actors work together with scientists and other stakeholders (governmental, non-governmental, etc.) to address locally relevant problems or issues. In fact, there is evidence of the potential of participatory approaches for addressing climate change issues in varied contexts (e.g. McClymont Peace & Myers, 2012; Middendorf et al., 2020). An example of this can be found in Canada, where the government developed a Climate Change and Health Adaptation Program for Northern First Nation and Inuit Communities with the intention of funding community-based participatory research for developing culturally appropriate and locally based adaptation strategies to reduce the effects of climate change on human health. Through this programme, the communities' understanding of the health effects of climate change has increased, and local adaptation strategies have begun to be developed (McClymont Peace & Myers, 2012).

Our case study in Piriápolis (coastal Uruguay), although not oriented to address climate issues, shows that the PAR process, contributed to developing the adaptive capacity of small-scale fishers in an integral manner, since the five domains identified by Cinner et al. (2018) were positively impacted, namely assets, flexibility, social organization, learning and agency. This is noteworthy since the five domains of adaptive capacity are necessary, and there appears to be limited substitutability between them with respect to shocks and long-term change (Cinner et al., 2018). Contributions focused mainly on the development of the social and human capitals. Also, our findings provided some examples of how progress achieved by the PAR process relates to the climate change vulnerability indicators proposed in the area (Teruggi, 2019), namely social integration, access to credit and insurance and diversification of production. We also observe that, having scaled-deep (i.e. "impacting cultural roots", changing relationships, cultural values and beliefs, "hearts and minds") (Moore et al., 2015) in the fishers' community of Piriápolis, the POPA group created by the PAR could be exerting still today, in an implicit manner, a positive influence in the different domains of the adaptive capacity of the fishers of Piriápolis to climate change vulnerability.

Similar to the Canadian case with Northern and Inuit communities (McClymont Peace & Myers, 2012), our case study highlights the active role played by the community actors (fishers) throughout all the phases of the developed projects (which were meaningful to them), along with scientists and other actors. In this regard, process evaluation criteria for participatory research (Trimble & Lázaro, 2014) include that the problem to be addressed must be of key interest to local and additional stakeholders; there should be involvement of interested stakeholder groups in every research stage, collective decision making through deliberation, and adaptability through iterative cycles of planning, acting, observing and reflecting. All of these should be considered as guidelines to develop empowering participatory research, which would allow for the achievement of outcomes such as knowledge co-production, learning, strengthened social networks and conflict resolution (Trimble & Lázaro, 2014), all of which are relevant for climate change adaptation. In conclusion, academics and practitioners should be encouraged to implement this approach for addressing climate change issues at the local scale and building climate resilience.

References

Adger, W. N. (2006). Vulnerability. *Global Environmental Change, 16*, 268–281.

Cheung, W. W. L., Lam, V. W. Y., Sarmiento, J. L., Kearney, K., Watson, R., Zeller, D., & Pauly, D. (2010). Large-scale redistribution of maximum fisheries catch potential in the global ocean under climate change. *Global Change Biology, 16*, 24–35.

Cinner, J. E., Adger, W. N., Allison, E. H., Barnes, M. L., Brown, K., Cohen, P. J., Gelcich, S., Hicks, C. C., Hughes, T. P., Lau, J., & Marshall, N. A. (2018). Building adaptive capacity to climate change in tropical coastal communities. *Nature Climate Change, 8*, 117–123.

Coulthard, S. (2012). Can we be both resilient and well, and what choices do people have? Incorporating agency into the resilience debate from a fisheries perspective. *Ecology and Society, 17*(1), 4. https://doi.org/10.5751/ES-04483-170104.

FAO. (2020). *El estado mundial de la pesca y la acuicultura 2020*. La sostenibilidad en acción. Roma. https://doi.org/10.4060/ca9229es.

Franco, B. C., Defeo, O., Piola, A. R., Barreiro, M., Yang, H., Ortega, L., Gianelli, I., Castello, J. P., Vera, C., Buratti, C., & Pájaro, M. (2020).

Climate change impacts on the atmospheric circulation, ocean, and fisheries in the southwest South Atlantic Ocean: A review. *Climatic Change, 162*, 2359–2377.

Gianelli, I., Ortega, L., & Defeo, O. (2019a). Modeling short-term fishing dynamics in a small-scale intertidal shellfishery. *Fisheries Research, 209*, 242–250.

Gianelli, I., Ortega, L., Marín, Y., Piola, A. R., & Defeo, O. (2019b). Evidence of ocean warming in Uruguay's fisheries landings: The mean temperature of the catch approach. *Marine Ecology Progress Series, 625*, 115–125. https://doi.org/10.3354/meps13035.

IPCC. (2014). *Climate change 2014: Synthesis report. Contribution of Working Groups I, II and III to the Fifth Assessment Report of the Intergovernmental Panel on Climate Change* [Core Writing Team, R.K. Pachauri and L.A. Meyer (eds.)]. (p. 151). IPCC, Geneva, Switzerland.

McClymont Peace, D., & Myers, E. (2012). Community-based participatory process—Climate change and health adaptation program for Northern first nations and Inuit in Canada. *International Journal of Circumpolar Health, 71*(1). https://doi.org/10.3402/ijch.v71i0.18412.

McGoodwin, J. (2001). *Understanding the cultures of fishing communities—A key to fisheries management and food security* (FAO Fisheries Technical Paper No. 401). Food and Agriculture Organization.

Middendorf, B. J., Vara Prasad, P. V., & Pierzynski, G. M. (2020). Setting research priorities for tackling climate change. *Journal of Experimental Botany, 71*(2), 480–489. https://doi.org/10.1093/jxb/erz360.

Moore, M., Riddell, D., & Vocisano, D. (2015). Scaling out, scaling up, scaling deep strategies of non-profits in advancing systemic social innovation. *The Journal of Corporate Citizenship, 58*, 67–85.

Pandey, R., Kumar Jha, S., Alatalo, J. M., Archie, K. M., & Gupta, A. K. (2017). Sustainable livelihood framework-based indicators for assessing climate change vulnerability and adaptation for Himalayan communities. *Ecological Indicators, 79*, 338–346.

Pretty, J., & Ward, H. (2001). Social capital and the environment. *World Development, 29*(2), 209–227.

Santos, P. (2017). *Un ejercicio de capacidad de voz: El caso de la Primera Feria de la Pesca Artesanal en Piriápolis (Uruguay)* (Master thesis in Communication and Culture). Universidad Católica del Uruguay, Montevideo.

Santos, P., & Trimble, M. (2018). Communication and culture for development: Contributions to small-scale fishers' wellbeing in coastal Uruguay.

In J. Servaes (Ed.), *Handbook of communication for development and social change*. Springer. https://doi.org/10.1007/978-981-10-7035-8_97-1.

Santos, P., Trimble, M., & Johnson, D. (2021). Balancing hope and disappointment: Representation, social wellbeing, and the future of small-scale fisheries in Uruguay. *Development in Practice*, 31(5), 580-591. https://doi.org/10.1080/09614524.2021.1907532.

Song, M. (2018). How to capture small-scale fisheries' many contributions to society?—Introducing the 'value-contribution matrix' and applying it to the case of a swimming crab fishery in South Korea. In D. Johnson (Ed.), *Social wellbeing and the values of small-scale fisheries* (pp. 125–146). Springer.

Sumaila, U. R., Cheung, W. W. L., Lam, V. W. Y., Pauly, D., & Herrick, S. (2011). Climate change impacts on the biophysics and economics of world fisheries. *Nature Climate Change, 1*, 449–456.

Teruggi, M. (2019). *A framework for assessing vulnerability to climate change of artisanal coastal fisheries in Uruguay*. University of Torino.

Trimble, M. (2013). *Towards adaptive co-management of artisanal fisheries in coastal Uruguay: Analysis of barriers and opportunities, with comparisons to Paraty (Brazil)* (Doctoral dissertation). University of Manitoba, Winnipeg.

Trimble, M., & Berkes, F. (2013). Participatory research towards co-management: Lessons from artisanal fisheries in coastal Uruguay. *Journal of Environmental Management, 128*, 768–778.

Trimble, M., & Johnson, D. (2013). Artisanal fishing as an undesirable way of life? The implications for governance of fishers' wellbeing aspirations in coastal Uruguay and southeastern Brazil. *Marine Policy, 37*, 37–44.

Trimble, M., & Lázaro, M. (2014). Evaluation criteria for participatory research: Insights from coastal Uruguay. *Journal of Environmental Management, 54*(1), 122–137.

Trimble, M., Iribarne, P., & Lázaro, M. (2014). Una investigación participativa en la costa uruguaya: características, desafíos y oportunidades para la enseñanza universitaria. *Desenvolvimento e Meio Ambiente, 32*, 101–117.

Weeratunge, N., Béné, C., Siriwardane, R., Charles, A., Johnson, D., Allison, E. H., Nayak, P. K., & Badjeck, M. C. (2014). Small-scale fisheries through the wellbeing lens. *Fish & Fisheries, 15*, 255–279.

Williges, K., Mechler, R., Bowyer, P., & Balkovic, J. (2017). Towards an assessment of adaptive capacity of the European agricultural sector to droughts. *Climate Services, 7*, 47–63.

Micaela Trimble holds a Doctoral degree in Natural Resources and Environmental Management from the University of Manitoba (Canada). She is an Associate at the South American Institute for Resilience and Sustainability Studies (SARAS), in Uruguay. She is a member of the National System of Researchers. Her areas of expertise include environmental governance and adaptive co-management of social-ecological systems, such as small-scale fisheries and watersheds.

Paula Santos holds a Master's degree in Communication and Culture at Universidad Católica del Uruguay. She is currently enrolled in a Master Programme on Economic History at the Faculty of Social Sciences of Universidad de la República, Uruguay. Her fields of study combine social well-being and resilience in socio-environmental vulnerable contexts. She is a Programme Assistant at the Regional Bureau for Sciences of UNESCO in Latin America and the Caribbean, Montevideo Office.

3

Ensuring Water Security for Climate Vulnerable Communities in Coastal Bangladesh

Istiakh Ahmed, Saleemul Huq, Shahrin Mannan, Md Juel Mahmud, and Mizan R. Khan

Abstract This chapter exposes the case of two coastal villages in Bangladesh, where climate change is likely to increase river salinity leading to shortages of drinking water and irrigation system at the coastal belt. Approximately 20 million people live along the coastline

I. Ahmed (✉) · S. Huq · S. Mannan · M. J. Mahmud · M. R. Khan
International Centre for Climate Change and Development (ICCCAD),
Dhaka, Bangladesh
e-mail: istiakh.ahmed@icccad.org

S. Huq
e-mail: saleemul.huq@icccad.org

S. Mannan
e-mail: shahrin.mannan@icccad.org

M. J. Mahmud
e-mail: juel.mahmud@icccad.org

M. R. Khan
e-mail: mizan.khan@icccad.org

© The Author(s), under exclusive license to Springer Nature
Switzerland AG 2021
P. H. Campello Torres and P. R. Jacobi (eds.), *Towards a just climate
change resilience*, Palgrave Studies in Climate Resilient Societies,
https://doi.org/10.1007/978-3-030-81622-3_3

and frequently rely primarily on rain-fed pond water and mix with river water, soil run-off and shallow groundwater. The study identifies specific vulnerabilities from different groups in these coastal villages. It examines the complex interlinkage of water security and explores how different environmental stressors and other factors negatively affect water security. Based on the findings, it also proposes recommendations to be made both to decision-makers at the national level to develop policies and plans to enhance the resilience of coastal communities.

Keywords Water security · Vulnerable communities · Bangladesh · Coastal areas · Climate change

Introduction

Located on an active delta, Bangladesh has always been a country with a high degree of environmental volatility, which poses high exposure to climatic hazards such as floods, cyclones, riverbank and coastal erosion (Mutahara et al., 2017). Researchers predict that the most critical impact of climate-induced disasters will be on the freshwater resources (IUCN). According to the Intergovernmental Panel on Climate Change (IPCC), groundwater and many rivers in coastal regions are likely to become increasingly saline from higher tidal waves and storm surges because of climate change impacts. Similarly, in Bangladesh, climate change is likely to increase river salinity leading to shortages of drinking water and irrigation systems at the coastal belt (World Bank Report, 2016).

Studies have suggested that the saline prone areas are likely to be gradually increasing and salinity will move further inland (Dasgupta et al., 2015; Miah, 2010). Models predicted that by 2050 an additional 15% of the coastal area of Bangladesh would be inundated with storm surges during cyclones. Changes in temperature and precipitation, combined with changes in the frequency and intensity of extreme hydro meteorological events, have widespread implications for water resources that affect the supply, quality and distribution of water resources for dwellers of these areas (Kundzewicz et al., 2007). In normal scenarios, the level of salinity at the coastal belt increases in the dry season and with heavy rainfall, it decreases during monsoon. Sea-level rise triggers salinity intrusion but it varies depending on the upstream freshwater flow through different rivers. During summer, more glaciers are melted in the Himalayas causing more precipitation and more entrance of freshwater. But during winter, much of that glacial ice remains frozen resulting in less freshwater flow from upstream and more saltwater from the bay intruding upstream into the delta. Furthermore, Tropical cyclones and tidal surges also exacerbate salinity in the region. Both after cyclone Sidr and Aila, river and land salinity increased significantly.

The coastal zone of Bangladesh covers 32% (47,201 km^2) area of the country, being the landmass of 19 districts (Ahmad, 2019) Increased River salinity coupled with prolonged dry season and lower River discharge in the face of a changing climate will lead to shortages of drinking water by 2050 (Naser et al., 2020). Approximately 20 million people living along the coastline of Bangladesh depend heavily on various natural sources to obtain water for drinking and domestic chores. Apart from River and groundwater (tube wells), people at the south-west coastal region frequently rely on pond water which are primarily rain-fed, but also mix with River water, soil run-off and shallow ground-water (Khan et al., 2011). A growing body of evidence already warns about the impact of increased dependency on saline water on human health and well-being; it is likely to worsen in the face of a changing climate. Expected impacts of climate change on water resources will be more pronounced due to poor infrastructure and fragile socio-economic structure (Dasgupta et al., 2014; Hossain et al., 2011).

This study intends to identify specific vulnerabilities experienced by different groups in two coastal villages, which share similar geographic and socio-economic characteristics of the larger portion of the coastal region. This book chapter examines the complex interlinkage of water security and explores how different environmental stressors and other factors are negatively affecting water security in coastal Bangladesh. Based on the findings, the chapter also proposes recommendations to be made both to decision-makers at the national level for development of policies and plans to enhance the resilience of coastal communities. A combination of quantitative and qualitative approaches was employed to guide research and data analysis, and the methodology can be replicated for undertaking similar kinds of detailed assessment for other locations and settings.

Research Sites and Methods

Overview of the Sites

The study was undertaken in two villages situated within the project's intervention area—Garuikhali village and Kumkhali village of Garuikhali Union in Paikgacha Upazila, Khulna. The sites chosen demonstrate similar geographic and socio-economic characteristics as the ones in Bagerhat district as well as other districts situated in the coastal belt. Literature review suggests that prevalent livelihood practices as well as water security issues present in these study sites are similar to the majority of the south-western coastal belt, affected by the same climatic shocks and stresses. Furthermore, coastal districts are characterized by similar poverty levels and issues of social inequity as observed in the two study sites. Findings of the study therefore provide an approximate snapshot of climate change vulnerability experienced by the majority of the coastal belt and solutions recommended can be replicated across the coastal region. However, unique socio-economic and environmental characteristics would need to be accounted for when prescribing ways forward for the other areas in the region.

Data Collection Tools

To address the research questions set by the study, a combination of top-down and bottom-up data collection approaches has been applied. Top-down approaches aimed to provide a scientific analysis of climate change and its impacts on the system of interest and included review of relevant literature as well as GIS modelling. Literature review helped understand the current status of vulnerability to climate change in the coastal region and prevalent gaps that exist. A snapshot of change projection for the region as well as future vulnerability was also drawn from the review. GIS modelling was applied to assess topographical changes in the region and how these changes relate to current and future exposure to climate change impacts.

Also, bottom-up approaches were employed to draw out necessary information from affected communities and to help analyse what causes people to be vulnerable to different hazards in the study area. An array of Participatory Rural Appraisal (PRA) tool (problem tree, transect walk, seasonal calendar, solution tree, participatory scenario analysis, community mapping, livelihood shocks, role playing exercise) was applied to understand key vulnerabilities of local communities, how they perceive risks to their water security and to identify strategies for mitigating these risks. A household survey was undertaken to acquire quantitative information on the overall socio-economic situation of the sites and a general view of exposure, sensitivity and adaptive capacity of the community to climate change. In addition, key informant interviews with relevant local stakeholders were held during the brainstorming phase of the study.

Sampling Procedure

Community discussion sessions for applying the different PRA tools, were held in selected parts of the villages. In selecting the respondents, several factors were considered such as diversity in live and different economic status. Separate group sessions for male and female respondents for each PRA tool. The individual interview respondents were selected for group discussions. The Key Informant Interview (KII) respondents were mostly relevant local government and non-government stakeholders as well as community people with a significant role in decision-making in the village.

Water Security and Interlinking Factors

Water security refers to the ability to access sufficient quantities of clean water crucial to maintain adequate standards of food and goods production, proper sanitation and sustainable health care (Dasgupta et al.,

2014). It exhibits four intersecting risks—environmental, institutional, financial and social. The environmental risks on coastal deltaic floodplains of Bangladesh are two-folded natural and human made. The natural form of environmental risks includes degradation of water quality with high levels of salinity and arsenic. Tidal flooding, over-extraction of groundwater coupled with reduced upstream flows are important natural causes of increased salinization (Tehsin & Mondal, 2017). The institutional risks largely comprise of uncoordinated policy making and service delivery, lack of accountability and poor management of water resources and infrastructure. Financial risks include shortfall in investments and low cost recovery (Tehsin & Mondal, 2017). Such combined risks further lead to a range of social risks including gender and wealth disproportionate inequalities affecting water security mental and physical distress and consequent health implications (Tehsin & Mondal, 2017).

Access to Water in Garoikhali and Kumkhali Village

Increased incidence of different climatic shocks and stresses in the villages pose significant threats to water security for local communities. Primarily, increasing levels of salinity across different water resources in the villages mean people are subjected to an array of water-related issues. Household survey revealed that in Garuikhali, for around 80% of the population, ponds serve as the primary source of drinking water, while the rest rely on rainwater. However, among the 80%, only about 5% are able to avail water from the only freshwater pond available in the village, which is more than 1 km away. These are mainly people who drive auto-rickshaws and can travel this long distance. They collect the water and end up selling them to other villagers. Majority of the population are unable to afford the water and end up consuming water from other saline sources. In Kumkhali, on the other hand, there are several tube wells as well as ponds that serve as sources of drinking water for the villagers. 58, 29 and 13% respondents reported ponds, tube wells and rainwater respectively as their primary sources of drinking water. Most households in Garuikhali have ponds alongside their homestead, while the number of ponds is comparatively less in Kumkhali. These ponds have long been major sources of freshwater but respondents in both the villages reported gradually increasing levels of salinity in these ponds over the past several years, particularly after the incidence of Cyclone Aila in 2009, which resulted in a substantial surge of saline water intrusion into these water bodies. Respondents in Kumkhali also reported increased salinity levels in tube wells than in the past. These observed changes in salinity levels were largely attributed to more frequent occurrence of cyclones as well as increasing heights of tidal surges, by respondents in both the villages. In addition, high-intensity, short-duration rainfall in the area often results in inundation of large sections of the two villages, including latrines. Inundation for extended periods of time means that impure water from these latrines intrudes into freshwater sources. Mobility for accessing freshwater sources is also restricted as a result of inundation. Freshwater access is further hindered by pollution of water bodies. A significant number of people in both the villages wash their clothes, cattle, dishes, take baths in the same pond where they also collect their drinking water from. For about 57.5% of respondents in Garuikhali and 40.5% in Kumkhali, the source of water for household activities was found to be the same as their drinking water source.

With the sources of potable and freshwater becoming increasingly scarce over last 5–7 years in Garuikhali, physical labour required for collecting water has also increased. For more than 90% of the respondents in the survey, the primary source of drinking water was found to be between 0–1 km away from their households, while only about 9% had to travel more than 1 km. Due to homestead ponds in Kumkhali possessing lower levels of the salinity, 70% of the respondents have to travel less than half a kilometre to collect water for drinking. See Graphs 3.1 and 3.2.

Women were found to be disproportionately affected in this regard as well. Traditionally, in rural Bangladesh, women are largely responsible for fetching drinking water from these sources and were found to be so in about 55.2% cases in Garuikhali and 73.5% cases in Kumkhali. Women have to travel the distance once or twice every day. Despite the availability of a number of tube wells in Kumkhali, they have mostly become dysfunctional. Time required for fetching water was also observed to have increased over the years. About 58.2% respondents in Garuikhali and

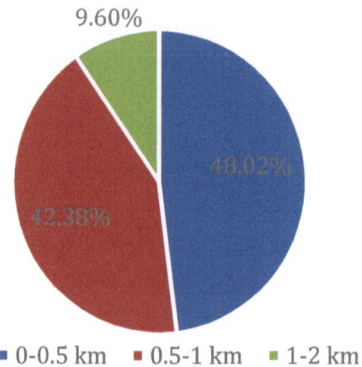

Graph 3.1 Distance to the main source of drinking water in Garuikhali

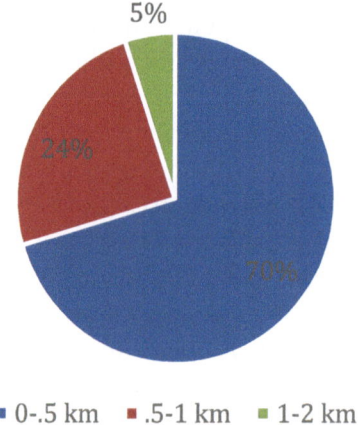

Graph 3.2 Distance to the main source of drinking water in Kumkhali

38.2% in Kumkhali, now require about half an hour to collect water from their source.

Causes of Water Scarcity and Its Effects in a Changing Climate

Increased incidences of different climatic shocks and stresses have a significant impact on the availability of potable water from both surface and groundwater sources in coastal areas. Figure 3.3 depicts a range of climate change non-climatic causes of water insecurity in the south-western coastal area.

Causes of Water Scarcity

High salinity in surface water sources is one of the major reasons of water insecurity. Proximity of this region to a number of rivers, such as Shibsa, Kabotak and Kholpetua, which go through seasonal fluctuation in the discharge, has been attributed as a key factor driving the severity

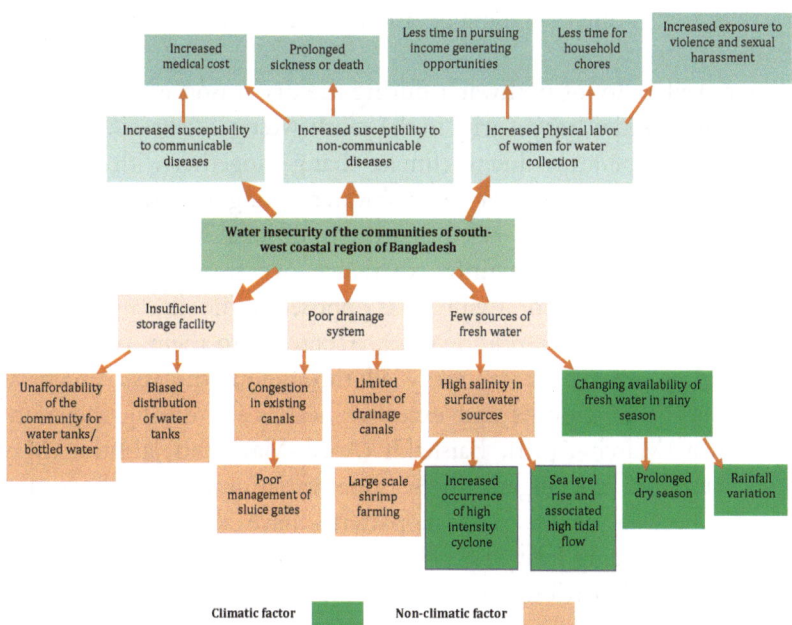

Fig. 3.3 Cause-effect relationship of water insecurities due to climate change

of storm surges associated with cyclone occurrence. With climate change induced sea-level rise and associated tidal surge, the southern coastal part of Bangladesh has already started facing intense coastal flooding together with increased intrusion of saline water in rivers and aquifers, irrespective of land protection. While some degree of salinity has historically always prevailed in the villages of south-western coastal region, a gradual increase in their level with the consecutive and increasingly frequent occurrence of cyclones over the last 30–40 years has been observed. Local communities blame cyclone Aila responsible for substantial saline water intrusion in both surface and groundwater sources. However, the salinity level tends to be high during the dry, summer months of Falgun (February–March), Chaitra (March–April), Baishakh (April–May) and Chaitra (May–June), and gradually decreases with the advent of rainy season in the months of Ashar (June–July) and Srabon (July–August). Owing to the prevailing saline conditions in this region, there has been an emergence of shrimp farming in the area, particularly by local elites, due to its high profitability. A rapid rise in large-scale shrimp farming practices is also one of the major driving forces behind increasing levels of salinity.

Poor availability of freshwater during the dry season is another driver for compromised water security in the south-west coastal region. Gradual increase in temperature due to climate change, together with changes in rainfall pattern lead to scarcity of freshwater during dry season. Community consultations have revealed that the traditional seasonal pattern has changed and now they mainly experience two seasons—warm season (summer) and cool season (winter), as opposed to the six-seasons climate Bangladesh has long been known for. According to them, temperatures now tend to remain warm nearly all throughout the year except for the extreme cold months and excessive heat is felt during the month of Chaitra (March–April), Baishakh (April–May) and Jaishtha (May–June). Moreover, a change in the seasonal variation of rainfall over time has also been observed. Rainfall patterns have also altered with the villages experiencing heavy rainfall over a short duration, as opposed to a more balanced distribution of rainfall over the monsoon months. Such high-intensity, short-duration results inundation and water-logging are affecting everyone. As a result of prolonged dry season and changes in rainfall pattern, communities face acute shortage of freshwater during dry season.

The issue of freshwater access in the two villages is further exacerbated by the lack of adequate water storage facilities available to local communities. To offset the problem of water scarcity during dry summer months, people tend to collect rainwater during the rainy season and try to preserve them as long as possible. However, they often run out of water too soon due to insufficient provisions for storing water. Graphs 3.4 and 3.5 delineate the state of having water reservoirs in the two villages.

Unfortunately, none of the water reservoirs and tanks that were installed by NGOs and the local government are functional currently. Furthermore, inadequate drainage mechanisms, supplemented by poor operation of sluice gates tend to prolong inundation and water-logging following days of heavy rainfall, thereby further contaminating water bodies in the area.

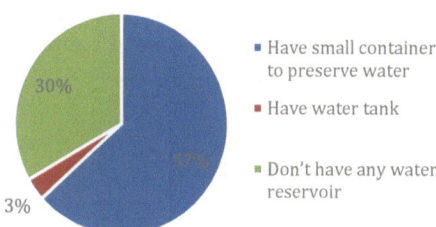

Graph 3.4 Percentage of people in Garuikhali with water storage facilities

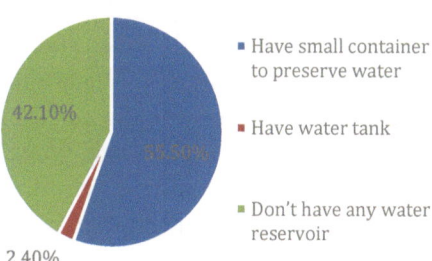

Graph 3.5 Percentage of people in Kumkhali with water storage facilities

Most of the people here have problem of storing fresh water/rain water as they don't have a reservoir/ water tank to store water. Only some of the effluent people who are politically affiliated with UP chairman, members have water tanks at their home. (Male, Village calendar, Garuikhali 19 July 2018)

Different non-climatic institutional, financial and social factors further reduce access to water, people's affordability and reliability of water systems, causing long-term water insecurity. Poor drainage system, particularly during the times of storm surges and coastal flooding lead to prolonged inundation and water-logging. Limited number of drainage canals and poor management of sluice gates, mostly operated in favour of shrimp farming, further exacerbate the challenges.

Access to freshwater is further exacerbated due to lack of adequate water storage facilities available to local communities. Local communities tend to collect rainwater during the rainy season to deal with water scarcity during summer months. However, most households neither have the affordability to buy water reservoirs/tanks nor bottle water. The distribution of water tanks by local NGOs and international donors are largely biased towards their enlisted members.

Effects of Scarcity of Freshwater

During the participatory scenario analysis and role-play exercise undertaken with the local communities, respondents primarily expressed concern regarding levels of salinity becoming worse in the coming years, further threatening water security in the area. Higher levels of salinity would mean higher incidence of health issues and diseases among the population. This would be accompanied by higher medical costs. Also, scarcity of nearby freshwater sources would lead to increased time and effort, especially from the women, in collecting water, thereby reducing productive hours. As a result, respondents felt that people who are already poor, would become poorer over time and overall socio-economic condition of the village will deteriorate as the impacts of climate change take precedence.

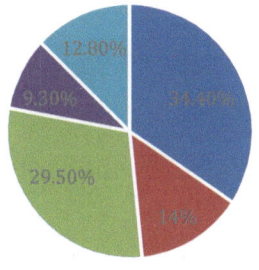

_{■ Women ■ Men ■ Children ■ Elderly ■ Disabled}

Graph 3.6 Percentage of disease affliction by population groups in Garuikhali

According to the respondents in both the villages, consumption and household usage of highly saline and contaminated water results in different water-borne diseases, such as diarrhea, cholera, stomach bugs and skin diseases. Women and children[1] are the ones most afflicted by these diseases. Cases of pregnancy-related problems such as hypertension during pregnancy and premature birth of children have been observed in recent years, which the respondents attribute to saline water consumption. Rise in malnutrition among children was also reported in Kumkhali. Graphs 3.6 and 3.7 demonstrate the distribution of these diseases across different demographic groups.

Furthermore, different non-communicable gastro-intestinal diseases and pregnancy-related complications, such as hypertension during pregnancy and also premature birth of children, have been observed in recent years, attributed to saline water consumption. All these compounding effects lead to increased medical cost and overall well-being loss of coastal communities. With the sources of potable and freshwater becoming increasingly scarce over the last 5–7 years, physical labour required for collecting water particularly among women has also increased. Travelling such long distances everyday leads to physical exhaustion and also occupies a substantial amount of time which could have gone towards other productive uses. Travelling such long distances everyday, leads to physical

---

[1] Children refers to persons below 15 years. According to UNICEF, youth refers to people ranging from 15–24 years; so any person below 15 years have been considered as children.

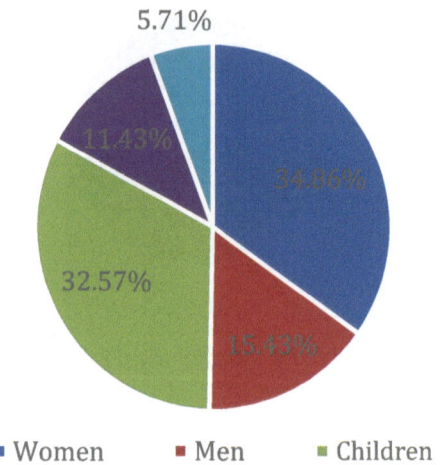

5.71%

11.43%

34.86%

32.57%

15.43%

■ Women   ■ Men   ■ Children

Graph 3.7 Percentage of disease affliction by population groups in Kumkhali

exhaustion and also occupies a substantial amount of time which could have gone towards other productive uses.

> Sometimes we have to go far away to collect potable water and it causes physical stress and sometimes illness to us. If I can save the time from fetching water, I can give more time in tailoring. (Mahmuda, Livelihood History, Garuikhali, 20 July 2018)

Furthermore, spending such long time fetching water from distant sources increases the change of facing sexual harassment.

Current Coping Strategies and Their Limitations

Though most of the water sources of coastal areas are not available for drinking as well as agriculture; government, NGOs and local community have been introducing different types of water technology to reduce the water scarcity. Some of them only work in the monsoon, some of them only remove turbidity and microbes, not saline and some of them are highly expensive. With limited resources, lack of maintenance and ownership, improper distribution these technologies are not serving efficiently for long run.

Technology	Activities	Limitation
Tube well (Shallow and Deep)	Extract groundwater from shallow depth Extract groundwater from deep aquifer	Most of the shallow aquifer contaminated with saline Arsenic contamination in most of the shallow aquifer Depletion of water table in dry season
Rainwater Harvesting (RWH)	Collect water from rain in the monsoon and storage for further uses	Irregular rainfall and long dry season (8–9 months) Lack of storage facilities
Pond Sand Filter (PSF)	Low cost, efficient filtration technology Remove turbidity and bacteria from pond water and make it usable	Due to lack of ownership and maintenance community-based PSF are not functioning for long time Cannot remove saline
Reverse Osmosis (RO)	Desalinized both surface and groundwater and produce freshwater	High installation cost Need energy source

Ensuring Water Security: How Community People Thinks!!

Living within the context, local people understand the problem better and thus can think of a solution that suits the best and matches with the need of the majority. During the role playing exercise, community people suggested their own way towards resilience in terms of water security.

- **Rainwater harvesting for each household**

People in both villages mostly drink water from a few selected ponds and in the very few tube wells present in Kumkhali. Since salinity is expected to persist in the villages, harvesting rainwater appears to be a feasible approach to ensuring access to freshwater. While the idea is not anything new, there are only few rainwater harvesting tanks in both the villages, and most of these do not have sufficient capacity to sustain water for long periods of time. Local government bodies and different NGOs have provided a few large sized tanks to selected households to be availed by everyone in the village. But in many cases, the host house does not allow others to use the water from these tanks.

Respondents therefore expressed that they will be benefited if each household in the village could receive a water tank. In addition to ensuring access for everyone, this will also reduce the potential for inter-community conflict over water use. Distribution of water tanks should also be based on needs.

The initiative could be taken up by some of the NGOs present in the area. Households also expressed willingness to invest some of their money for establishing these tanks. Respondents mentioned that the plastic water tanks are better than concrete ones, as soil salinity tends to damage the concrete tanks and plastic ones are also cheaper. Presently, plastic rainwater harvesting tanks cost about BDT 3000–4000, and each family is willing to contribute 1000–2000 taka towards that. Adequate availability and access to freshwater can ensure good health and also save time for collecting water and enhancing productivity, thereby enhancing the overall well-being of people.

- **Desalinization Plant at Community Level**

To combat the issue of salinity in both drinking water as well as water for other usage, a desalinization plant can be set up in the community. There are some examples of community level desalinization plants established in the nearby Satkhira district by different NGOs and few of the respondents have had experience of consuming water from such filters.

- **Freshwater irrigation system**

There is no proper irrigation system in these villages. In Kumkhali, they have to draw water from the pond or tube well and use it for irrigation. While the entire area is prone to salinity, a freshwater canal from where they can do irrigation for cultivating their lands would be a good solution for improving agricultural production and promoting livelihood empowerment. Shrimp farming is not feasible for the largely poor population and hence a robust, freshwater irrigation system can allow people to grow crops and vegetables to supplement their incomes. Respondents suggested a proper irrigation system that can serve as a pipeline distribution of water from a definite source or pump system, which is currently unavailable in their village.

- **Proper management of sluice gate**

According to the respondents, the sluice gate should be operated and managed properly by the local government. The closing and opening time should be maintained properly. There is no person in charge of operating sluice gates at the moment and for this reason, they have to suffer from unwanted saline water intrusion. Saline water should also be drained out from the area on time so that saline water inundation would not occur. In addition to proper management of sluice gates, respondents highlighted the need for more responsible use of water. This includes reducing wastage and refraining from contaminating water bodies by littering waste. This will help reduce health-related impacts in the village.

- **Canal dredging**

Kumkhali village has a canal and they believe it is very unlikely to get another canal nearby. If this canal can be properly managed, it can provide freshwater irrigation for the entire area. But the canal has never been dredged in its entire period and over the time has filled up a lot. At this rate, it is expected to dry up soon and respondents have already observed that its depth has reduced in the last 20 years. Respondents emphasized the need to dredge the canal to ensure freshwater flow and reduce water scarcity.

Conclusion

This research study sheds light on the current scenario of the coastal region of Bangladesh and possible future climate change consequences. Being at such close proximity to the sea and rivers, salinity intrusion is one major stressor affecting the local livelihood pattern, which once was fully based on agriculture but now has shifted to shrimp farming, in most of the places. However, shrimp farming has also contributed to further soil salinity affecting nearby agricultural fields and reducing harvest yields. Increased salinity levels in groundwater tables have also caused severe drinking water scarcity within the locality and currently there is only one pond for a fairly large community to access drinking water.

While many poor people are now unable to cultivate their land (mainly due to soil salinity) and shrimp farming is not labour-intensive (does not need large numbers of labour force to operate), this has triggered seasonal migration from this region to nearby cities across Bangladesh. Based on the results of the study, it is evident that climate change induced shocks and stresses are already manifesting themselves in an array of adverse impacts on the social and economic well-being of communities in the two coastal villages. The study aimed to explore how local communities perceive changes in the climate, as observed by higher incidence and severity of climatic shocks and stresses, and iden-tified resultant impacts on access to water resources as well as overall well-being. Differentiated effects of these changes on different groups and the interlinkages among these impacts have also been recognized.

The study shows that climate change exacerbates vulnerabilities caused by non-climatic factors (such as shrimp farming, poor water resource management and weak governance structures). If the present envi-ronmental and socio-political situations persist, water security of local communities is likely to substantially deteriorate, further worsening their state of well-being. To help build resilience of these communities, there is a need to look deeper into the current absorptive and adaptive capacity of vulnerable communities in the study area—by exploring in detail, different indicators such as knowledge base, asset base, infrastructure availability, natural resource base, access to health care, social capital and governance structures.

References

Ahmad, H. (2019). Bangladesh coastal zone management status and future. *Journal of Coastal Zone Management, 22*(1). https://doi.org/10.24105/2473-3350.22.466.

Dasgupta, S., Kamal, F. A, Khan, Z. H., Choudhury, S., & Nishat, A. (2014). River salinity and climate change. In *World scientific reference on Asia and the world economy* (pp. 205–242).

Dasgupta, S., Hossain, M. M., Huq, M., & Wheeler, D. (2015). Climate change and soil salinity: The case of coastal Bangladesh. *Ambio*. https://doi.org/10.1007/s13280-015-0681-5

Hossain, M. A., Reza, M. I., Rahman, S., & Kayes, I. (2011). Climate change and its impacts on the livelihoods of the vulnerable people in the Southwestern coastal zone in Bangladesh. In *Climate change management* (pp. 237–259).

IUCN. (n.d.). *Climate change and water in Bangladesh*. Retrieved from IUCN website: https://www.iucn.org/sites/dev/files/import/downloads/water.pdf.

Khan, A., Ireson, A., Kovats, S., Mojumder, S., Khusru, A., Rahman, A. R., & Vineis, P. (2011). Drinking water salinity and maternal health in coastal Bangladesh: Implications of climate change. *Environmental Heath Perspectives, 119*, 1328–1332.

Kundzewicz, Z. W., Mata, L. J., Arnell, N. W., Döll, P., Kabat, P., Jiménez, B., Miller, K.A., Oki, T., Sen, Z., & Shiklomanov, I. A. (2007). Freshwater resources and their management. In M. L. Parry, O. F. Canziani, J. P. Palutikof, P. J. van der Linden, & C. E. Hanson (Eds.), *Climate change 2007: Impacts, adaptation and vulnerability. Contribution of Working Group II to the fourth assessment report of the Intergovernmental Panel on Climate Change* (pp. 173–210). Cambridge University Press.

Miah, M. M. U. (2010). *Assessing long-term impacts and vulnerabilities on crop production due to climate change in the coastal areas of Bangladesh*

(Final Report PR #10/08). http://fpmu.gov.bd/agridrupal/sites/default/files/ Muslem_Uddin_Miah-PR10-08.pdf.

Mutahara, M., Warner, J.F., Wals, A. E. J., Khan, M. S. A., & Wester, P. (2017). Social learning for adaptive delta management: Tidal River Management in the Bangladesh Delta. *International Journal of Water Resources Development*, 1–21.

Naser, A., Dosa, S., Rahman, M., Unicomb, L., Ahmed, K., Anand, S., & Selim, S. (2020). Consequences of access to water from managed aquifer recharge systems for blood pressure and proteinuria in south-west coastal Bangladesh: A stepped-wedge cluster-randomized trial. *International Journal of Epidemiology, 1*, 1–13. https://doi.org/10.1093/ije/dyaa098

Tehsin, S., & Mondal, M. (2017). *Assessing agricultural water security for different agro-ecosystems in a coastal area of Bangladesh using analytical hierarchy process*. Paper presented at 6th International Conference on Water and Flood Management.

World Bank Group. (2016). Impact of climate change and aquatic salinization on fish habitats and poor communities in Southwest coastal Bangladesh and Bangladesh Sundarbans.

4

Policy and Planning Responses to Climate Change in Solomon Islands: A Place for Forest-Based Carbon Offset Initiatives in Building Just and Resilient Territories?

Kristen Lyons and Peter Walters

Abstract Pacific Island nations are on the front line of some of the worst climate change impacts. This chapter takes the case of the Solomon Islands to examine local livelihood, policy and planning responses to climate change. While a negligible per capita greenhouse gas emitter, Solomon Islands national government, and alongside civil society and others are enacting a range of climate change responses. Amongst these includes a number of forest-based carbon offset initiatives. While these projects are enabled by conducive national and international policy and carbon market settings, their impetus is grounded in an alliance of tribal leaders, local community organisations and environmental NGOs committed to balancing conservation with positive livelihood outcomes.

K. Lyons (✉) · P. Walters
School of Social Science, University of Queensland, Brisbane, QLD, Australia
e-mail: kristen.lyons@uq.edu.au

P. Walters
e-mail: p.walters@uq.edu.au

© The Author(s), under exclusive license to Springer Nature
Switzerland AG 2021
P. H. Campello Torres and P. R. Jacobi (eds.), *Towards a just climate
change resilience*, Palgrave Studies in Climate Resilient Societies,
https://doi.org/10.1007/978-3-030-81622-3_4

Drawing lessons from Choiseul Province, this chapter assesses the place of forest-based carbon offset initiatives in building more resilient territories, including those that centre both ecologies and people in responding to the climate crisis.

Keywords Solomon Islands · Planning responses · Forest carbon offset · Resilience · Environmental justice

Introduction

There are diverse global responses to the urgent challenge of a changing climate. For those on the frontlines of the climate crisis, including Pacific and other small states, the need for urgent action to address the worst impacts is unambiguous (Alston, 2014; Lyons, 2019). In Solomon Islands—the focus of this chapter—climate change presents acute ecological, social, cultural and economic challenges. Sea-level and temperature rise threaten settlements, subsistence agriculture and fishing and exacerbate already vulnerable forest, marine and other unique and biodiverse ecosystems (Albert et al., 2016).

Reflecting these dangers, Pacific Island leaders have described climate change as the "single greatest threat to the livelihoods, security and well-being of the peoples of the Pacific" (Leannem, 2018, p. n.p.). The Small Islands Development States Group (SIDS) and the Alliance of Small Island States (ASIS) have called for urgent global action to ensure the just transition to a zero carbon economy (Evans & Phelan, 2016). Pacific Nations have also taken the lead in a Call for Action that ensures climate change responses are grounded in principles of democratic participation, science, social justice, ethical leadership and cooperation (Suliman et al., 2019). These principles will be vital, Pacific Nation leaders assert, to ensure "sustainable development and the preservation of life on earth as we know it" (Banos Ruiz, 2018, n.p.).

Taking Solomon Islands as case study, this chapter critically assesses some of the policy and planning responses to the challenge of climate change, as well as opportunities for these responses to enable environmentally just and resilient territories in the face of the current realities of global carbon markets. To do this, our chapter begins with a brief introduction to Solomon Islands, including current environmental challenges, with a particular focus on logging given its centrality in national greenhouse gas emissions. We then introduce some of the key adaptation and mitigation policy and planning responses to climate change, including forest-based carbon offset; one of a number of rapidly expanding climate mitigation initiatives. Carbon offset is now widely supported, including on the basis of its promise to provide a pathway for decarbonisation that avoids major disruptions to sectors and industries (including energy, transport, food and agriculture, etc.), as well as generating massive profits (Lang, 2020).[1]

We then introduce our environmental justice approach, which we apply to analyse Solomon Islands' policy and planning responses to climate change. Our analysis highlights some of the opportunities and challenges for forest-based carbon offset initiatives to contribute to climate change responses to ensure just and resilient societies. We conclude that commitment to both distributive and procedural forms of justice will be vital if carbon offset initiatives are to be part of a climate just future. To achieve this, local communities must be involved in decision-making that affects them, including in ways that empower them to directly shape benefits and other outcomes. Recognition of local and Indigenous rights over land and water will also be fundamental to achieving any benefits.

Background: Solomon Islands, Development and Forests

Solomon Islands, with a population of 670,000 (UN, 2019), is a small Melanesian island state in the Pacific Ocean to the South East of Papua New Guinea. A British colony for 88 years, Solomon Islands achieved political independence in 1978. The archipelago consists of six

---

[1] Carbon offset relies on activities in one location that sequester—or absorb—carbon dioxide and other greenhouse gas emissions in another location. For many heavy polluting western countries, carbon offset enables industries and sectors to maintain rates of greenhouse gas pollution, by offsetting their emissions at another site (e.g. forests, forestry plantations, etc.).

major, and 900 smaller, islands. The population is spread through the archipelago and is predominantly rural, with a concentration of 86,000 in the capital, Honiara, on the island of Guadalcanal. Ninety-two per cent of the population adheres to a variety of Christian denominations. Solomon Islands was ranked 153/189 on the UNDP Human Development Index in 2019 (UNDP, 2019). The domestic economy is supported predominantly by subsistence agriculture and fishing. Foreign exchange through exports relies overwhelmingly on extractive industries, mainly tropical hardwood timber and, more recently, mining (Allen & Porter, 2016).

The most contentious of these industries to date has been the large-scale selective logging of primary rainforests by mainly Malaysian logging companies, although the rapid expansion of mining now generates a range of environmental and human rights issues and concerns (see, e.g., Allen & Porter, 2016). Logging for export has expanded rapidly on native-owned lands since independence and coinciding with growing international demand (Bennett, 2000; Dauvergne, 1998/1999). For the past 25 years, logging has taken place at a rate of 250,000 m^3 per year; over 20 times what is widely recognised as a sustainable rate (Allen, 2011). Forests and logs are accessed by sea and an estimated 12,500 km network of purpose-built logging roads (Katovai et al., 2015). Soil runoff into the sea caused by damage to riparian systems from logging, roads and log ponds has created severe ecological damage to reefs and traditional fishing grounds (Hamilton et al., 2017). While it has been estimated that by 2036 accessible timber will have been exhausted, logging continues as companies negotiate new agreements with traditional landowners to extract timber in areas where trees have matured since the first logging.

Logging is notionally controlled by a system of government issued licences in consultation with traditional landowners; however, corruption is rife. Logging has been enabled by systemic fraud, corruption and misconduct, including the sustained direct involvement of politicians in the logging industry (Allen, 2011). While Solomon Islands' largest logging operator—Malaysian owned Rimbunan Hijau—was officially removed from the country in April 2020 (Sarawak Report, 2020), a messy and opaque structure enables its continued operation, albeit under different trading names (Oakland Institute, n.d.a, b). Such conditions generate selective incentives for inequitable short-term wealth that causes ongoing intra-tribal tensions and outcomes that continue to divide communities and drive adverse social and environmental impacts (Allen, 2018; Oakland Institute, n.d.a, b).

Forests—and logging—are now central to debates related to the challenge of climate change in Solomon Islands, and elsewhere. The urgency of forest management was reflected in a recent IPCC (2019) report that noted agriculture, forestry and other land uses accounted for 23% of global greenhouse gas emissions. Deforestation was singled out as a key contributor; with Duggin (n.d.) reporting that logging, including illegal logging, responsible for significant total greenhouse gas emissions. On this basis, land use policies continue to have a critical impact on the climate (IPCC, 2019). This is particularly so in Solomon Islands. While there is no nationally available data, ongoing logging in Solomon Islands makes it one of the highest emitting countries in the Pacific[2] (Ministry of Environment, Climate Change, Disaster Management and Meteorology, 2017).

Following this brief background to Solomon Islands' environmental issues, we now describe some of the key climate change challenges facing the region.

Climate Change and the Pacific

The effects of current climate change in small islands states of the Pacific[3] are well known and have been described as acting "as a spatial prism through which injustices and inequities inherent to anthropogenic climate change are reflected and some of its worst effects realised" (Suliman et al., 2019, p. 299). Throughout the Pacific, tides are rising,

---

[2] It should also be noted that Solomon Islands contributes just 0.17 Metric Tonnes of CO_2 per year compared with the US: 5189 Mt CO_2/year and China: 10,358 Mt CO_2/year (Kumar et al., 2020, p. 12).

[3] We use the term Pacific, and Pacific Island nations, to assist in setting out some of the generalised impacts associated with climate change for the region. In so doing, we acknowledge that impacts are localised and diverse and that various communities, regions and countries will be impacted in different ways (Barnett & Campbell, 2010).

villages are being more regularly inundated and some relocations are beginning (Kumar et al., 2020). Rising temperatures, changes in rainfall patterns and levels of ocean acidity are affecting agricultural practices, traditional fisheries and coral bleaching; diminishing local food sovereignty and increasing the incidence of disease (Hanich et al., 2018). Climate change impacts also threaten the culture and traditions of Indigenous populations, including destroying sites of cultural significance (Kumar et al., 2020). There is also growing evidence of the gender-based impacts of climate change, including increasing workloads for women including as labour and livelihoods shift in the face of a changing climate, alongside a rise in gender-based violence caused by disruption to traditional livelihoods and cultural practises (McLeod et al., 2018).

While the effects of climate vulnerability might be predominantly physical (e.g. sea-level rise), their underlying causes are tied to economic, social and political contexts, and reflect colonial legacies that continue to drive adverse outcomes for Indigenous communities (Cardona, 2004; Whyte, 2017). In contrast to dominant narratives that paint Pacific Island nations as victims of climate change however, Indigenous and local communities in the Pacific are also on the frontlines of reimagining understandings of, and responses to, the challenges of climate change.

Demonstrating this, Indigenous and local communities have taken the lead in bringing local knowledges to climate change responses, including through the IPCC and other policy and planning responses (see, e.g., Rashidi & Lyons, 2021). Meanwhile the denial of rights in the face of a changing climate—which must be upheld as set out by the United Nations Declaration on the Rights of Indigenous Peoples (UNDRIP)— is driving a new wave of Indigenous-led climate litigation (Barnett & Campbell, 2010; Lyons et al., 2021). Similarly, in the face of forced migration, Indigenous communities are reimagining "ways of moving in a warming world" (Suliman et al., 2019, p. 299). As example, the Pacific Climate Warriors—an alliance of Pacific Island activists—demonstrate forms of mobilisation and resistance that centre Indigenous epistemologies, knowledges and rights as the basis for reimagining the future. Their declaration: "we are not drowning we are fighting" speaks, for example, to their defence of an enduring right to land and sea (Goodyear-Kaopua, 2017). In the context of a broader rights of nature project, Pacific Island communities are leading a campaign to secure legal recognition for the rights of the Pacific Ocean. This is based on enduring understandings of the Ocean as so much more than just water or a food store for Pacific Islanders; "it has its mana (spiritual authority) and *mauri* (life force) (David, 2019, p. 1).

An Environmental Justice Approach

In this chapter, we adopt an environmental justice framework to assess some of the policy and planning responses to the challenge of climate change in Solomon Islands. With its attention on balancing environmental and social dimensions, including a global and human rights focus, environmental justice provides a lens to assess the ways Solomon Islands' climate change responses might foster just and resilient territories (McCauley & Heffron, 2018).

Environmental justice approaches the issue of justice from diverse perspectives; including a focus on distribution, participation and recognition (see, e.g., Holifield et al., 2018; Schlosberg, 2007). Distributive justice, firstly, draws attention to the ways in which diverse actors are able to enjoy the environmental, economic and other benefits of nature and ecosystems (and so-called 'natural resources'), and to avoid environmental risks and harms. In the context of a changing climate, distributive justice examines proximity to climate-related risks and harms, including sea-level rise, as well as rights and access to nature (including forests, forest 'resources' and so on). While ethnicity and race are of central significance—demonstrated via the disproportionate environmental risks and harms facing communities of colour—newly emerging distributional frameworks also identify the significance of capabilities and well-being in shaping local-level experiences of a changing climate (McCauley & Heffron, 2018). In so doing, this approach considers how the distribution of climate change impacts is shaped by a diverse set of intersecting socio-cultural, economic and other factors.

Secondly, procedural justice refers to the ways in which decisions about environmental management, including rights and access to nature are made. This draws attention to people's responsibilities in decision-making, as well as the rules governing these processes (Schlosberg, 2007). Procedural justice examines processes related to participation, including the ways in which local communities might be engaged in ways that open up broadly inclusive opportunities for globally and locally led decision-making (McCauley & Heffron, 2018). Research in this space has exposed some of the limits for effective participation, including local communities' insecure property rights, lack of financial resources and alongside weak institutions as each constraining procedural justice (Iftikhar et al., 2007).

Finally, justice as recognition draws attention to the groups that are respected in decision making processes, and the basis upon which such validation is accorded. This approach draws attention to the political and cultural status accorded to different identity groups (including based upon gender, ethnicity, and so on) and situates this within histories of colonialism, racism, and other forms of discrimination. The failure to recognise Indigenous peoples' rights to, and knowledge of, biological and ecological systems, as example, exposes both a failure in recognition and the structural racism inherent in dominant conservation approaches. Justice as recognition thereby exposes the ways diverse forms of injustice are interconnected, and with outcomes that disavow local and Indigenous knowledges, culture and governance systems (see Martin et al., 2020).

In the context of a changing climate, environmental justice—including distributive, procedural and recognition justice—provides an approach to examine rights and interests in relation to climate change policy and planning. By centring consideration on fairness and equity in efforts to shift to a post carbon society, our chapter thereby contributes to the broader just transitions literature (see, e.g., McCauley & Heffron, 2018).

Climate Change and Solomon Islands' Policy and Planning Responses

Solomon Islands has engaged substantially with international organisations related to global climate policies and planning. This is translated in diverse ways at the regional and local levels. Globally, Solomon Islands is Party to the United Nations Framework Convention on Climate Change (UNFCCC) and its Kyoto Protocol, which together make up the core of international policy responses to climate change. As signatory, Solomon Islands has joined global efforts to curb greenhouse gas emissions to stabilise temperature rise at no more than 2° above pre-industrial levels. It is also signatory to the Hyogo Framework on Disaster Risk Management and has been involved in the European Union-Global Climate Change Alliance programmes. Solomon Islands also receives funding from the Global Environment Facility (GEF) (the financing mechanism for the UNFCCC), made available through the UNDP, UNEP, FAO and World Bank (Ministry of Culture, Tourism & Aviation, 2015).

At the regional level, Solomon Islands is signatory to a number of plans and agreements to establish climate change and disaster risk management objectives and actions, including the 1Pacific Plan, Pacific Islands Framework for Action on Climate Change (PIFACC) and the Regional Framework on Disaster Risk Reduction and Disaster Management (Ministry of Culture, Tourism & Aviation, 2015). At the national level, Solomon Islands' overarching development planning framework is aligned with Solomon Islands National Development Strategy: 2011–2020 (NDS). Solomon Islands has also implemented a range of climate adaptation and mitigation[4] strategies, informed by its National Adaptation Program of Action, National Disaster Management Strategy and Renewable Energy Frameworks.

---

[4] The intergovernmental Panel on Climate Change (IPCC) defines adaptation as "the process of adjustment to actual or expected climate and its effects". Quite simply, it means adapting (food, transport, urban design, energy, etc.) to live with a changing climate (Noble et al., 2014). Meanwhile, they define mitigation as "human intervention to reduce the sources or enhance the sinks of greenhouse gases" (IPCC, 2014), and can include replacing heavy emitting energy sources with renewables, introducing energies efficiencies across buildings and industries, as well as tree planting for greenhouse gas absorption and storage.

Within these international, regional and national policy and planning contexts, Solomon Islands' government has initiated a number of climate policy-related directives. These include seeking to mainstream climate change as core business across all areas of government, as well as identifying vulnerability, adaptation and disaster risk reduction as key focus areas. It has also initiated a number of forest-based carbon mitigation directives. These are aimed at strengthening the capacity for design and implementation of Clean Development Mechanism (CDM) projects.[5] Solomon Islands government has sought to strengthen governance and capacity for carbon trade through the UN Reducing Emissions from Deforestation and Forest Degradation (REDD+) programme and Voluntary Carbon Trading and via establishing carbon trading legislation (Ministry of Culture, Tourism & Aviation, 2015).

Turning to REDD+, the Solomon Islands Government, in collaboration with NGOs, provincial governments and community leaders, has produced a REDD+ Roadmap and formally adopted REDD type policies and programmes as part of its national climate change adaption and mitigation policy agenda. This policy context has occurred alongside an expansion in the certification and registration of certified sustainable forest management initiatives and forest conservation projects (Corrin, 2012). This shift towards REDD+ signals the most recent development in Solomon Islands' forest governance policies. While Solomon Islands' forests were historically worth economically more felled, the global reconceptualisation of forests as repositories of carbon is driving shifts in forest management (Wairiu, 2007; Walters & Lyons, 2016).

Within this policy and planning context, forest-based carbon offset projects are now underway in Solomon Islands. These initiatives provide insights about the place of forest-based carbon offset in supporting a

---

[5] The Clean Development Mechanism (CDM)—as detailed in Article 12 of the Kyoto Protocol—allows countries with a commitment to reduce emissions to implement emission reducing activities in a (so-called) developing country, including Solomon Islands (see https://cdm.unfccc.int/index.html). Nearly 3000 CDM projects have been implemented in Asia and Pacific under the Kyoto Protocol, representing 82% of all projects. Most of this project funding has been directed to China and India, with far less allocated to countries most vulnerable, such as Sri Lanka, Bangladesh, as well as Pacific nations (Nakhooda, 2013).

just climate future. We now examine these projects through an environmental justice lens.[6]

Forests as Carbon and a Just Transition?

On the Island of Choiseul at the Western extremity of Solomon Islands' archipelago, there are a small number of current projects designed to protect native forests, while integrated with international carbon markets (see Lyons et al., 2019). These initiatives are managed by a Solomon Islands based, internationally funded NGO—the Natural Resources Development Foundation (NRDF)—on sites where logging has not been extensive. The NRDF project is a collaboration with the Nakau programme (a Pacific-wide forest conservation NGO) and is based on a logic of forest preservation and carbon trading for the benefit of Indigenous landowners (Nakau, 2017).

In Choiseul, NRDF is working with tribal communities to achieve two aims; the first is compliance with Forest Stewardship Council (FSC) certification and carbon credit programmes to ensure communities meet the external criteria levied on them by certification organisations as the basis for entry into carbon markets; and second, to implement a range of livelihood programmes to either complement income from carbon credits, or to act as alternative sources of income should carbon credit income fail to materialise. The purpose of both these programmes is to provide tribal communities with real alternatives to allowing logging companies onto their land for short-term financial gain and long-term environmental damage.

Programmes initiated by NGOs such as NRDF and enabled by national and international carbon policy settings as detailed above, may bring financial and conservation benefits to traditional landowners. Yet

---

[6] It is beyond the scope of this chapter to assess the environmental justice implications of international, regional and national climate policy and planning settings in which Solomon Islands' forest-based carbon offset projects are implemented. These themes are taken up elsewhere (see, e.g., Comberti et al., 2016; Rashidi & Lyons, 2021 forthcoming), with research demonstrating carbon politics is vexed, with small island states (including Solomon Islands) and Indigenous peoples frequently sidelined and/or excluded.

there are issues of concern in integrating local level actors into a new frontier of global capital manifest in the use of existing sources of carbon capture to offset the unsustainable practises of corporate actors in heavily polluting sectors elsewhere in the world.

There is an ongoing critique of carbon markets in general and REDD+ in particular. In terms of distributive justice, carbon market mechanisms are opaque to all but those with an invested expertise in these mechanisms. This expertise is generally held by those operating at the global level and often applied as a result of the international commodification and trade of carbon. As a result, decisions on carbon market transactions are determined by market mechanisms radically beyond the control or knowledge of forest custodians in Solomon Islands. This lack of control, knowledge and information means traditional forest owners will always be price-takers, with little ability or certainty for planning or investing carbon offset income in the medium or long term.

There are also impacts for procedural and recognition justice. Local communities and traditional owners are far from equal partners in carbon markets, despite their ownership of valuable carbon resources. Traditional owners live and work in local economies that are deeply socially and culturally embedded. Carbon markets are, by their nature, global and abstract. Carbon offset is a strategy designed to prolong tangible carbon emissions and delay its reduction. By enabling big industrial polluters such as power generators and airlines to continue emitting through offsetting emissions against already-existing natural carbon resources, traditional owners are recruited as unwitting accomplices to this environmental procrastination strategy. There is also evidence that carbon offset projects are susceptible to deception, fraud and bad faith using sophisticated globalised financial instruments and a lack of international regulation and oversight to ensure offset schemes are achieving the goals for which they were established (Carrington, 2010). There are also debates about whether offset initiatives actually reduce emissions in the atmosphere (see Carbon Market Watch, 2013).

Böhm et al. (2012) argue that the institution of global markets, rather than pivoting to a greener, and a more climate sensitive form of global capitalism, is merely an extension of existing unjust global power relations. This raises other questions of recognition justice and how global carbon offset practices and communities such as those in Solomon Islands might be integrated into global carbon markets in ways that further extend extractivist colonial power relations. Although carbon credits do not impose the obvious local destruction that logging and mining do, the extraction of carbon from one environment makes the Solomon Islands inadvertently complicit in the continuation of environmental degradation in other locations on the planet.

We should also note that local impacts of the carbon market are not all negative. In Choiseul province, the local carbon projects as described above (Lyons et al., 2019), are managed by a locally embedded NGO with sufficient foresight and, perhaps, scepticism about the long-term sustainability and reliability of carbon trading. In preparing local communities and their forests for the quantitative demands of participation in carbon markets, they have also, in parallel, worked to establish alternative livelihood options for communities, which are compatible with forest sustainability and allow for the failure of carbon market income to materialise. Initiatives such as beekeeping, organic food production and women's saving and local microcredit facilities have all allowed for modest income generating potential in the absence of carbon offset income. These allow communities a certain level of resilience against the short-term temptations of opening their land to logging operations. The methodology applied by organisations such as NRDF provides a valuable blueprint for local level collaborative interventions in other small state contexts.

Conclusions

Drawing lessons from Solomon Islands carbon offset initiatives, this chapter has assessed the place of forest-based carbon offset initiatives in building just and resilient territories, and a just transition to climate resilient environments and economies in small island states. We conclude that distributive and procedural justice—where local communities are able to directly realise the benefits of these initiatives, including taking an active role in decision-making—are vital to global efforts to mitigate against the worst effects of climate change. The rights and interests of local communities to determine how their land and water resources are used is also critical if carbon offset initiatives are to be part of a climate just future.

There are larger questions about the legitimacy of forest-based carbon offsets in reducing emissions. The economics and politics of global carbon trade will continue to make equitable participation very hard to achieve as long as the current logic of economic growth and borderless capital flows are maintained.

How might Solomon Islands' climate change policy and planning responses resonate with distributive and procedural justice, as well as justice recognition? What lessons might be learnt from this assessment related to forest-based carbon offset initiatives in particular, in building just and resilient territories? Solomon Islands, in important ways, is an example of best practise for the implementation of carbon market projects in a small state. The government has fully engaged with international organisations, and at the local level, socially and culturally invested NGOs are providing a clear-eyed strategy for carbon projects, focussing on education, forest protection and alternative livelihoods in anticipation of uncertain returns from carbon markets. However, the fact remains that small states like Solomon Islands have few avenues to engage with, or influence, global capital. For now, that is where the true power lies in the carbon market.

References

Albert, S., Grinham, A., Gibbes, B., Leon, J., & Church, J. (2016). Sea-level rise has claimed five whole islands in the Pacific: First scientific evidence. *The Conversation*. Available at: https://theconversation.com/sea-level-rise-has-cla imed-five-whole-islands-in-the-pacific-first-scientific-evidence-58511.

Allen, M. (2011). The political economy of logging in Solomon Islands. In R. Duncan (Ed.), *The political economy of economic reform in the Pacific* (pp. 278–301). Asian Development Bank.

Allen, M. (2018). *Resource extraction and contentious states. Mining and the politics of scale in the Pacific Islands*. Palgrave Macmillan.

Allen, M., & Porter, D. (2016). Managing the transition from logging to mining in post-conflict Solomon Islands. *The Extractive Industries and Society, 3*(2), 350–358.

Alston, M. (2014). Gender mainstreaming and climate change. *Women's Studies International Forum, 47*(Part B): 287–294.

Banos Ruiz, I. (2018). *Talanoa dialogue: Giving everyone a voice in the climate conversation.* DW. Available at: https://www.dw.com/en/talanoa-dialogue-giv ing-everyone-a-voice-in-the-climate-conversation/a-42479711.

Barnett, J., & Campbell, J. (2010). *Climate change and small island states: Power, knowledge, and the South Pacific.* Earthscan.

Bennett, J. A. (2000). *Pacific forest: A history of resource control and conflict in Solomon conflict.* Brill Academic Publishers.

Böhm, S., Misoczky, M. C., & Moog, S. (2012). Greening capitalism? A Marxist critique of carbon markets. *Organization Studies, 33*(11), 1617–1638.

Carbon Market Watch. (2013). REDD. Carbon Market Watch. https://carbon marketwatch.org/2013/04/09/redd/.

Cardona, O. (2004). The need for rethinking the concepts of vulnerability and risk from a holistic perspective: A necessary review and criticism for effective risk management. In G. Bankoff, G. Frerks, & D. Hilhorst (Eds.), *Mapping vulnerability: Disasters, development and people* (pp. 56–70). Routledge.

Carrington. (2010). *EU plans to clamp down on carbon trading scam.* https:// www.theguardian.com/environment/2010/oct/26/eu-ban-carbon-permits.

Comberti, C., Thornton, T., & Korodimou, M. (2016). *Addressing Indige-nous peoples' marginalisation at international climate negotiations: Adaptation and resilience at the margins* (Working Paper, ECI). University of Oxford. Available at https://papers.ssrn.com/sol3/papers.cfm?abstract_id=2870412.

Corrin, J. (2012) *Redd+ and forest carbon rights in Solomon Islands: Background legal analysis.* International Climate Initiative. Federal Republic of Germany: On behalf of Federal Ministry for the Environment, Nature Conservation and Nuclear Safety.

Dauvergne, P. (1998/1999). Corporate power in the forests of the Solomon Islands. *Pacific Affairs, 71*(4), 524–546.

David, V. (2019). Towards a regional convention on the rights of the Pacific Ocean as a legal entity. *One Ocean Symposium, New York.* Available at https://www.griffith.edu.au/__data/assets/pdf_file/0031/848722/Victor-David-Regional-Convention-on-Rights-of-the-Pacific-Ocean-as-a-legal-ent ity-24Aug19.pdf.

Duggin, G. (n.d.). *Climate change and "REDD": How the Solomon Islands' forests fit in the global response to climate change.* Environmental Defenders Office. Available at https://www.pacificclimatechange.net/sites/default/files/docume nts/EDO_redd_presentation_April_2010.pdf.

Evans, G., & Phelan, L. (2016). Transition to a post-carbon society: Linking environmental justice and just transition discourse. *Energy Policy, 99*, 329–339.

Goodyear-Kaopua, N. (2017). Protectors of the future, not protestors of the past: Indigenous Pacific activism and Mauna a Wakea. *South Atlantic Quarterly, 116*(1), 184–194.

Hamilton, R. J., Almany, G. R., Brown, C. J., Pita, J., Peterson, N. A., & Howard Choat, J. (2017). Logging degrades nursery habitat for an iconic coral reef fish. *Biological Conservation, 210*, 273–280. https://doi.org/10.1016/j.biocon.2017.04.024

Hanich, G., Wabnitz, C., Ota, Y., Amost, M., Donato-Hunt, C., & Hunt, A. (2018). Small scale fisheries under climate change in the Pacific Islands region. *Marine Policy, 88*, 279–284.

Holifield, R., Chakraborty, J., & Walker, G. (2018). Introduction: The worlds of environmental justice. In R. Holifield, J. Chakraborty, & G. Walker (Eds.), *The Routledge handbook of environmental justice*. Routledge.

Iftikhar, U., Kallesoe, M., Duraiappah, A., Sriskanthan, G., Poats S., & Swallow, B. (2007). *Exploring the inter-linkages among and between Compensation and Rewards for Ecosystem Services (CRES) and human well-being* (World Agroforestry Centre Working Paper 36), Nairobi, Kenya, pp. 1–44.

IPCC. (2014). Summary for policymakers. In O. Edenhofer et al. (Eds.), *Climate change 2014: Mitigation of climate change. Contribution of working group III to the fifth assessment report of the intergovernmental panel on climate change.* Cambridge University Press.

IPCC. (2019). *Land is a critical resource* (IPCC report says). In IPCC (Ed.). Geneva.

Katovai, E., Edwards, W., & Laurance, W. F. (2015). Dynamics of logging in Solomon Islands: The need for restoration and conservation alternatives. *Tropical Conservation Science, 8*, 718–731.

Kumar, L., Jayasinghe, S., Gopalakrishnan, T., & Nunn, P. D. (2020). Climate change and impacts in the Pacific. In (pp. 1–31). Springer International Publishing.

Lang, C. (2020). Mark Carney's taskforce on scaling voluntary carbon markets: The global financial Elite's plan to profit from the climate crisis while maintaining business as usual for big oil. *REDD Monitor.* Available at https://redd-monitor.org/2021/01/26/mark-carneys-taskforce-on-scaling-vol untary-carbon-markets-the-global-financial-elites-plan-to-profit-from-the-cli mate-crisis-while-maintaining-business-as-usual-for-big-oil/.

Leannem. (2018). Pacific Islands prepare for the UN climate negotiations, the battle for survival. *Pacific Environment.* Available at: https://www.sprep.org/news/pacific-islands-prepare-for-the-un-climate-negotiations-the-battle-for-survival.

Lyons, K. (2019, May 27). Torres Strait Islanders ask UN to hold Australia to account on climate human rights abuses. *The Conversation.* Available at https://theconversation.com/torres-strait-islanders-ask-un-to-hold-australia-to-account-on-climate-human-rights-abuses-117262.

Lyons, K., Esposito, A., & Johnson, M. (2021). The Pangolin and the coal mine: Challenging the forces of extractivism, human rights abuse and planetary calamity. *Antipode Intervention*

Lyons, K., Walters, P., & Shewring, A. (2019). *"Forests for Life"* or forests for carbon markets? The case of Choiscul Province, Solomon Islands *Pacific Dynamics. Journal of Interdisciplinary Research, 3*(1), 1–14.

Martin, A., Armijos, M. T., Coolsaet, B., Dawson, N., Edwards, G., Few, R., Gross-Camp, N., Rodriguez, I., Schroeder, H., Tebboth, M., & White, C. (2020). Environmental justice and transformations to sustainability. *Environment: Science and Policy for Sustainable Development, 62*(6). https://doi.org/10.1080/00139157.2020.1820294.

McCauley, D., & Heffron, R. (2018). Just transition: Integrating climate, energy and environmental justice. *Energy Policy, 119,* 1–7.

Ministry of Environment, Climate Change, Disaster Management and Meteorology. (2017). Solomon Islands Second National Communication to UNFCCC. Solomon Islands, Honiara. Nakau. (2017). The Nakau Program. R

McLeod, E., Arora-Jonsson, S., Masuda, Y., Bruton-Adams, M., Emausrois, C., Gorong, B., Hudlow, C., James, R., Kuhlken, H., Lasike-Liri, B., Musrasrik-Carl, E., Otzelberger, A., Relang, K., Reyuw, B., Sigrah, B., Stinnett, C., Tellei, J., & Whitford, L. (2018). Raising the voices of Pacific Island women to inform climate adaptation policies. *Marine Policy, 93*(July), 178–185.

Ministry of Culture, Tourism and Aviation. (2015). Solomon Islands. Initial national communications under the United Nations Framework Convention on Climate Change Report. Ministry of Culture, Tourism and Aviation, Honiara.

Nakau. (2017). *The Nakau Program.* http://www.nakau.org.

Nakhooda, S. (2013). *The effectiveness of international climate finance*. Overseas Development Institute.

Noble, I., Huq, S., Anokhin, Y., Carmin, J., Goudou, D., Lansigan, F., Osman-Elasha, B., & Villamizar, A. (2014). Adaptation needs and options. In C. Field et al. (Eds.), *Climate change 2014: Impacts, adaptation, and vulnerability. Part A: Global and sectoral aspects. Contribution of working group II to the fifth assessment report of the intergovernmental panel on climate change* (pp. 833–868). Cambridge University Press.

Oakland Institute. (n.d.a). *The great timber heist: The logging industry in Papua New Guinea*. The Oakland Institute.

Oakland Institute. (n.d.b). *The great timber heist—Continued: Tax evasion and illegal logging in Papua New Guinea*. The Oakland Institute.

Rashidi, P., & Lyons, K. (2021). How Democratic is global climate governance? The case of indigenous representation in the Intergovernmental Panel on Climate Change (IPCC). *Globalisations*. Special issue—Comparative politics of decarbonization: From carbon democracy to climate democracy? https://www.tandfonline.com/doi/full/10.1080/14747731.2021.1979718

Sarawak Report. (2020). Predator logging company thrown out of Solomon Islands is owned by sons of elected Sarawak YB—Expose! Available at https://www.sarawakreport.org/2020/01/predator-logging-company-thrown-out-of-solomon-islands-is-owned-by-sons-of-sarawak-mp-expose/.

Schlosberg, D. (2007). *Defining environmental justice: Theories, movements, and nature*. Oxford University Press.

Suliman, S., Farbotko, C., Ransan-Cooper, H., & Kitara, T. (2019). Indigenous (im)mobilities in the anthropocene. *Mobilities, 14*(3), 298–318.

UN. (2019). *World population prospects 2019*. Retrieved from Geneva: https://population.un.org/wpp/Download/Standard/Population/.

UNDP. (2019). *2019 Human development report*. Retrieved from New York: http://hdr.undp.org/en/content/2019-human-development-index-ranking.

Wairiu, M. (2007). Forest certification in Solomon Islands. *Yale School of Forestry and Environmental Studies*, 137–162.

Walters, P., & Lyons, K. (2016). Community teak forestry in Solomon Islands as donor development: When science meets culture. *Land Use Policy, 57*, 730–738.

Whyte, K. (2017). Indigenous climate change studies: Indigenising futures, decolonising the Anthropocene. *English Language Notes, 55*(1–2), 153–162.

5

Cabo Delgado, Mozambique: Beyond Climate—How to Approach Resilience in Extremely Vulnerable Territories?

Carla Gomes⊙ and Luísa Schmidt⊙

Abstract Northern Mozambique has been faced with increasing compound risks over the last few years, with rising climatic and political instability, in a region already affected by persistent poverty despite positive growth trends at the national level. Cyclone Kenneth and the COVID-19 pandemic, amid rising political violence, have afflicted Cabo Delgado populations and aggravated the situation of those most vulnerable, posing a very significant threat to the fulfilment of the Sustainable Development Goals (SDGs) in this region. This chapter analyses risks affecting this region from the perspective of human capabilities and just adaptation. It analyses how multiple vulnerability factors led to 'clusters'

C. Gomes (✉) · L. Schmidt
Institute of Social Sciences, University of Lisbon, Lisbon, Portugal
e-mail: carla.gomes@ics.ulisboa.pt

L. Schmidt
e-mail: mlschmidt@ics.ulisboa.pt

© The Author(s), under exclusive license to Springer Nature
Switzerland AG 2021
P. H. Campello Torres and P. R. Jacobi (eds.), *Towards a just climate change resilience*, Palgrave Studies in Climate Resilient Societies,
https://doi.org/10.1007/978-3-030-81622-3_5

of disadvantages over the last few years (2015–2021), which calls for an integrated approach to climatic and post-COVID resilience.

Keywords Mozambique · Vulnerability · SDG · Capabilities · Political instability

Northern Mozambique has been faced with increasing compound risks over the last few years, with rising climatic and political instability, in a region that was already affected by persistent poverty despite positive growth trends at the national level. Cyclone Kenneth and the COVID pandemic, amid rising political violence, have afflicted Cabo Delgado populations and aggravated the situation of those most vulnerable, posing a very significant threat to the fulfilment of the Sustainable Development Goals in this region.

Despite its remoteness and low levels of development, the northernmost province of Cabo Delgado (2500 kms from the capital Maputo) is rich in environmental and mineral resources—rubies were discovered in Montepuez in 2009, and liquid natural gas near Palma in 2010. This has attracted the interest of transnational investors such as the UK's Gemfields, France's Total and Italy's Eni. There has been though strong controversy over the distribution of the benefits amongst local populations, especially when considering negative impacts such as displacement of local communities to give way to the extractive industries. The unequal distribution of benefits from foreign investments and the lack of opportunities for the younger generations, in a context of persistent poverty, has been considered as breeding conditions for the expansion of Islamist group Al-Shabaab since October 2017.

In a world with a SDG agenda and facing a climate emergency, the question stands: How to address multidimensional vulnerability in territories where multiple pressures have been reinforcing each other and surpassing resilience tipping points? What direction to take under such extreme humanitarian situations?

There has been an extensive body of literature over the last couple of decades that transitioned from human ecology to human geography and other social sciences with the mission to better understand how different socio-ecological systems are able to respond, adapt and perhaps even thrive in the face of environmental change, most notably climate change. Resilience has been a key concept in explaining how ecological systems, but also societies, have transitioned and responded to multiple shocks.

It has become clear, in this debate, that social features such as social and family networks have been determinant amongst other factors to reinforcing the resilience of local actors to external factors such as hurricanes and other climate disasters (Adger & Kelly, 1999). However, other factors of vulnerability, such as political violence and pandemics, directly harm the ability to find support in these family and social networks, thus creating 'clusters' of disadvantage (Wolff & De-Shalit, 2007) and deprivation of human capabilities (Nussbaum, 2011; Sen, 2013).

With a view on clearing up the debate on a much used concept, Walker (2020) has defined the key attributes of resilience: response diversity; being modular—not over or under-connected; being able to respond quickly to shocks and changes in the system; being ready to transform if necessary; thinking, planning, and managing across scales; guiding, not steering. The perspective of resilience presupposes that, beyond a certain tipping point, environmental shocks require a deep transformation of human societies (Otto et al., 2017).

More recently, social movements have also brought to the fore new concepts that have become swiftly embedded into mainstream political discourse. We are said that we need to transition to a low-carbon society, but more than that we need to undergo a profound social transformation in order to also adapt to inevitable climate change and overall environmental collapse (Pelling et al., 2014).

What has been certainly a challenge is looking at environmental and social change from an integrated and interdisciplinary perspective. If we bring into the debate human development scholars we have to consider that environmental shocks will play into a world where most of the population is already deeply deprived across multiple dimensions. These multiple dimensions, or human capabilities, interact to form clustering disadvantages (Wolff & De-Shalit, 2007), or to reinforce each other as advantages.

We situate the debate at this crossroads between environmental change and human development in order to reach a broader understanding of the reinforcing disadvantages that have been plummeting the most vulnerable countries across the Global South. We will look into one of the poorest and climate vulnerable countries in the world, Mozambique, which despite a trend of economic growth over the last few decades has plummeted into a deprivation spiral over the last few years, under the compound pressure of climate disaster, economic crisis and political instability. The pandemics of COVID-19 is one more factor aggravating the disarray in this country, where climate disaster and terrorism had already caused mass displacement, especially in the North and central regions.

The country has been affected by recurrent tropical storms—Idai in March 2019, Kenneth six weeks later in April 2019, Chalane in December 2020, Eloise in January 2021. When Chalane hit the central region, it destroyed part of the resettlement camps that were still in place in the wake of Idai. In the North, Islamist insurrection provoked mass displacements and social instability in the aftermath of Kenneth's devastation and pushed thousands to crowded rescue shelters in the midst of the COVID-19 pandemics.

"The simplest definition of resilience is the ability to cope with shocks and to keep functioning in much the same kind of way. It is a measure of how much an ecosystem, a business, a society can change before it crosses a tipping point into some other kind of state that it then tends to stay in" (Walker, 2020). It seems that in just three years this thriving Southern African nation has crossed that tipping point. Furthermore, in face of multiple sources of vulnerability, we ought to pay attention to possible trade-offs, as building resilience to one kind of threat may hinder resilience to other situations. What does this mean, then, for the global development agenda, at such a critical time, most notably for the Sustainable Development Goals (SDGs)?

In this chapter, we look into the multiplying factors of vulnerability, at economic, social, environmental and political levels, that have affected Northern Mozambique since 2015, with a special focus on the situation of Cabo Delgado. We then proceed to analyse these multiple vulnerabilities as deprivation of human capabilities, towards an integrated approach to resilience and the fulfilment of the SDG in such extremely vulnerable territories.

Roots of Vulnerability in Northern Mozambique: The Case of Cabo Delgado (2015–2021)

Since the end of the civil war in 1992, Mozambique has been widely considered a donor's 'darling' and was fast climbing the ladder of the Human Development Index from the very last places in the world ranking (Cunguara & Hanlon, 2012). Although the national trend of economic growth has been upwards and relatively stable, there have been persistent disparities across the country. The Northern provinces have typically been lagging behind since the early 2000s compared to Southern, and of course more notably Maputo. It is widely acknowledged that benefits of economic growth have not been equally distributed, 'robust and relatively stable but not inclusive' (World Bank, 2018, p. 67).

Cabo Delgado, along with other Northern provinces such as Nampula and Sofala, ranks amongst the highest rates of chronic malnutrition in the country. Similar situation for sanitation (20%); education (70%); access to water and electricity. The poverty rate lies at over 50%. Investment in basic infrastructure and social services has been considered amongst the key factors hindering poverty reduction and human well-being over the last couple of decades.

Whilst growth in household consumption accelerated after 2008, lifting more people out of poverty, it is becoming progressively less inclusive, benefiting disproportionally more the better off. One of the reasons pointed for this has been the slow structural transformation as the economy transitions out of agriculture, as well as scarce opportunities for high-quality jobs, which are skewed towards urban, male and skilled workers (World Bank, 2018).

Raising productivity and resilience to shocks in agriculture, the main source of livelihood for most of the poor in the years to come (World Bank, 2018), and therefore, climate adaptation is undoubtedly at the forefront in overcoming the current humanitarian and poverty crisis in Mozambique.

Over the last few years, Mozambique has been through a succession of multiple shocks at the economic, political and environment level, which now hinder the fulfilment of the SDG in the country. On the wake of the global recession (2008), in 2015 the scandal of the hidden debt plummeted the country into a financial crisis of its own, which predictably hit the least well-off especially harshly. Followed by a succession of serious droughts and hurricanes (Idai and Kenneth just over six weeks in 2019), Islamist insurgency in the North since 2017, and since early 2020, along with the rest of the world, the COVID-19 pandemics, which is contributing to massive deprivation in convergence with other factors, such as pest outbreaks and weather shocks (Fig. 5.1).

Despite its remoteness, as we have mentioned above, Cabo Delgado has attracted the interest of sectors such as extractive industries and tourism. The latter was pointed out as the most promising opportunity for economic, social and cultural development for local people in the province of Cabo Delgado in the Poverty Reduction Strategy Paper (PARP, 2011–2014) (International Monetary Fund, 2014), as the capital

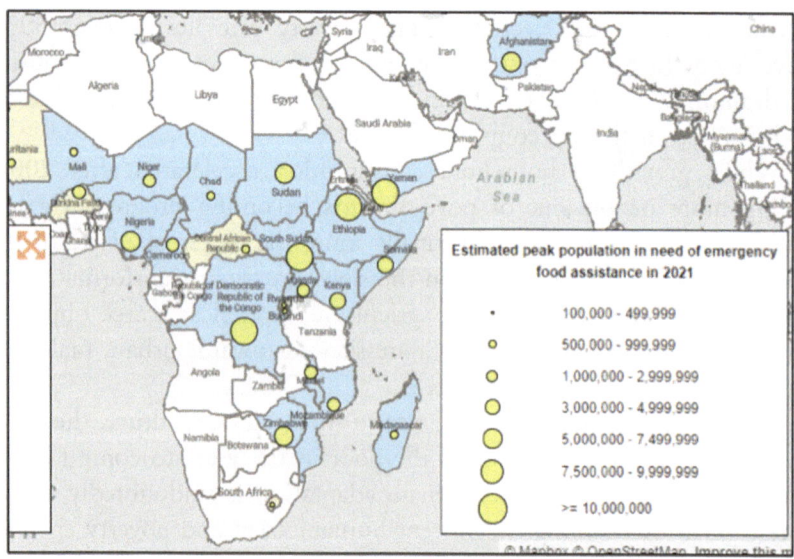

Fig. 5.1 COVID-19 Pandemic Impacts on Food Security (*Source* FEWS NET; https://fews.net/covid-19-pandemic-impacts-food-security)

Pemba has been the main jumping-off point for tourism in nearby Quirimbas National Park.

The Government has designed large-scale development programmes to be implemented in the deprived North, in the case of Cabo Delgado the Rio Lúrio Development Programme, set to establish multiple agricultural projects along the margins of the river, in the provinces of Niassa and Cabo Delgado. However, these have raised concerns amongst NGOs and researchers that implementation of large-scale agricultural projects will ultimately lead to rural displacement and disrupt the livelihoods and food security of smallholder families.

The gas reserves found in Northern Cabo Delgado along Palma are said to be amongst the largest in the world, and analysts estimate this could make Mozambique the third largest exporter globally of liquefied natural gas. Although extraction has not yet begun, disputes about occupation of the land have been intensifying, and most recently the Islamist insurgency is putting the gas extraction project on hold.

The maldistribution of benefits from tourism and extractive investments, amongst rising inequality and the added pressure of the 'hidden debt' crisis in 2015 paved the way for the current explosive situation of political violence in the 'Forgotten Cape' (Amnesty International, 2021).

Climate Events and Vulnerability

The agricultural sector employs nearly 80% of the labour force, which is one of the highest rates across countries in Sub-Saharan Africa, mostly smallholder rain-fed farming and extremely vulnerable to climate shocks. The FAO has pointed out the rain-fed maize-based agricultural systems of Sub-Saharan Africa amongst the most vulnerable in the world (FAO, 2017).

The 2015 flooding in Mozambique was a bigger than usual and mostly unexpected natural event that caused huge damage to infrastructure, especially roads and bridges, estimated at about 2.4% of GDP. Salvucci and Santos (2020) estimate that this single climatic event resulted in a drop of 11–17% in household consumption, affecting particularly the poorest families in rural areas, and an increase of 6% in poverty levels. Damage to transportation, communication, and electricity infrastructures was amongst the main causes for these significant social and economic impacts, which are very expressive of how climate vulnerability, per se, has the power to set back the fulfilment of the SDG. Just four years later, the same region was hit by cyclone Kenneth. Cyclone Kenneth was the strongest tropical cyclone ever to hit the African continent and left about 374,000 people in need.

Due to the combined impacts of the cyclones and drought in localised areas in the northern province of Cabo Delgado, an estimated 1.65 million people in Mozambique were assessed to be severely food insecure between the period of June and September 2019, according to the latest IPC analysis. This figure is projected to increase to nearly 2 million people facing Crisis or worse levels of acute food insecurity from October 2019 to February 2020 (as of September 2019) (USAID, 2019). Figure 5.2. presents estimations for 2021/2022.

One year after Tropical Cyclone Kenneth struck Cabo Delgado in Mozambique on 25 April 2019, more than 390,000 people had received critical assistance, such as food, clean water and shelter. Floods in December 2019 and January 2020 displaced thousands of people and damaged key infrastructure, exacerbating the needs of over 200,000 people who were still living in destroyed or damaged homes and another 6,600 people still sheltering in tents at the time (OCHA, 2020).

The country as a whole has been affected by increasing intense tropical storms. Tropical cyclone Dineo hit Mozambique in February 2017. According to the government, approximately 550,000 people were affected, and more than 33,000 homes were completely destroyed. The Idai (2019) and Chalane (2020) hit the central region. 4,938 families who were still in resettlement sites had their tents and shelters destroyed or partially destroyed as a result of Tropical Storm Chalane, which is a clear demonstration of how recurrent climate extremes are making it increasingly difficult for populations to recover, as well as for organisations to provide emergency response. Idai killed more than 600 people and left an estimated 1.85 million people in need. Tropical Cyclone Eloise, the strongest tropical storm since Kenneth, hit the central region again in January 2021. Nearly 250,000 people were affected, and nearly 17,000 houses impacted.

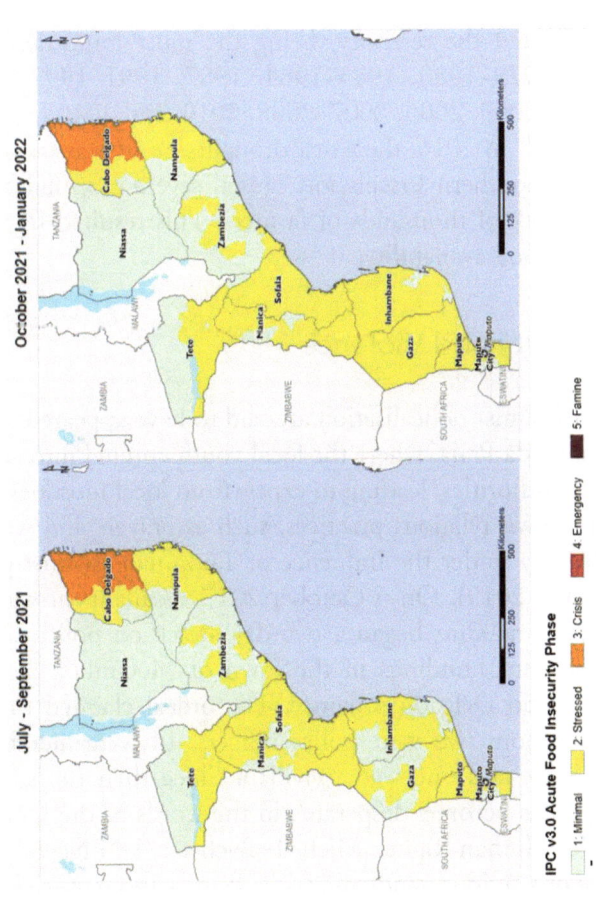

Fig. 5.2 "Food insecurity crisis in Northern Mozambique (estimates for July 2021 to January 2022)" (*Source* FEWS NET); https://fews.net/southern-africa/mozambique/key-message-update/july-2021/ (as of 25 August 2021)

Extreme climate events are therefore making it increasingly difficult to break the vicious cycle of crop loss and displacement, with vast percentages of the population requiring emergency food assistance annually.

But not just tropical storms have been increasingly depleting agriculture and depriving rural populations in Mozambique. Recurrent droughts have affected the territory, being the most important those that occurred in 1979–1980, 1983–1984, 1987, 1991–1992, 1994–1995, 1998, 2001–2003, 2005, 2007–2008, 2010, and 2016 (Araneda-Cabrera et al., 2021). In 2016, the worst droughts in 35 years associated with the El Niño–Southern Oscillation caused severe crop failure and the death of livestock of thousands of farmers. This resulted in a 15% decrease in overall food availability.

Political Instability and Violence

The first signs of Islamist radicalisation are said to have appeared around 2015 in Mocímboa da Praia, where the local youth entered into conflict with the Muslim authorities, leading to expels from local mosques. They soon started their own religious practices, such as prayers and wedding ceremonies, allegedly under the influence of Tanzanian Islamist groups (Matsinhe & Valoi, 2019). On 5 October 2017, assailants armed with machetes and machine guns began a two-day attack on police stations and other government buildings in the town of Mocímboa da Praia, in the north of Cabo Delgado province. The ordeal claimed the lives of 17 people, of whom two were police officers, 14 were members of the attacking group, and one was a civilian. Since then the situation has escalated and has become 'desperate', in the words of the UN High Commissioner for Human Rights, Michelle Bachelet (UN News, 2020).

In the middle of a serious pandemic crisis, Pemba and its surrounding areas, such as the nearby town of Metuge, have received a growing number of displaced people from the Northern part of the province. Many people move in with their extended families—often as many as 30 or 40 people into a single-family dwelling (Amnesty International, 2021)—or to crowded rescue camps, which raises the risks of contracting

COVID and has sparked a serious mental health crisis. The displaced people, as well as the local communities who welcomed them, have been under strong psychological strain. Although there is an atmosphere of fear and many people are afraid or too traumatised to speak, there are many reports of beheadings, executions, and women have been particularly affected, being frequently raped, sequestered and even forced to marry the insurgents (Ahmed, 2021). Older women and men have been particularly affected by the conflict, as well as persons with disabilities who were unable to flee fighting (Amnesty International, 2021).

It is estimated that 70% of the displaced are women and children, and there are reports of many still hidden in the woods, without access to food, water, or any assistance. In late 2020, interviews from Amnesty International revealed stories of strong deprivation and food insecurity. Displaced populations have naturally left most of their belongings, farmlands and even families behind and therefore are extremely vulnerable and dependent of emergency assistance, or help from acquaintances and family.

The Armed Conflict Location and Event Data Project (ACLED) estimated that over 1,300 civilians had been killed during the conflict in a total of 798 'organized violence events' in Cabo Delgado between October 2017 and February 2021, and the UN High Commissioner for Refugees estimated that more than half a million people—or over one quarter of the entire population of Cabo Delgado—were displaced internally.

Towards an Integrated Approach to Resilience and Human Capabilities

Northern Mozambique, and particularly Cabo Delgado, is a case that achingly demonstrates how the confluence of multiple factors of vulnerability can send populations and territories into a spiral of deprivation, surpassing the 'tipping points' of socio-ecological resilience and threatening human well-being to unsustainable levels. The notion of clustering disadvantages from Wolff and De-Shalit (2007) resonates well with this context, along with Nussbaum's central capabilities.

The recent and additional factors of vulnerability that we discussed in this chapter feed into a complex 'cluster of disadvantages', especially centred on poor and mostly rural populations, creating an especially difficult situation that hinders the fulfilment of the SDG since their creation in 2015. This refers mostly to SDG1 (No Poverty), SDG2 (Zero Hunger), SDG10 (Reduced Inequalities) and SDG13 (Climate Action).

Massive displacement is the core turning point for a spiralling process of vulnerability—due to political conflict, violence and climate disaster—that can be analysed from the lens of Nussbaum's central capabilities.

The speed of the change turns it difficult to secure adequate support to vulnerable populations on their **bodily health** and need for shelter. When a new hurricane hits, part of the population is still living on the improvised shelters from the previous one, which further increases their exposure and delays recovery. Islamist attacks force millions to cities such as Pemba, increasing pressure on the already vulnerable populations who welcome them, and on their scarce resources.

Needless to mention the threat to lives and bodily integrity in a context of extreme political violence prompted by the Islamist insurgency. In addition to the profound shock of losing loved ones and leaving relatives behind, as well as homes, lands, crops and belongings (**control over one's environment**).

The pandemics complicate the management of shelter for the displaced, raising risks of contagion and increasing pressure over resources generally. It also impedes the grieving of loved ones and the sharing of human **emotions**.

Recurrent climate extremes, including droughts and tropical storms such as Kenneth (2019) not only deepen food insecurity, increasing climate vulnerability jeopardises people's relation to **other species and nature** over the immediate and the longer term.

The convergence of factors of vulnerability at multiple scales, and most notably the climate emergency, requires, more than ever, an integrated socio-environmental approach, towards a transformative adaptation that fully addresses the root causes of vulnerability (Pelling et al., 2014).

On one hand, it is determinant that environmental policy, namely climate adaptation and the national plans for adaptation (NAPs), be harmonised with long-standing development cooperation in the country, including its institutions, NGOs, initiatives and funding programmes, based on robust socio-environmental integrated assessments that take into full account regional disparities, social inequity and justice.

On the other hand, it has become clear that this integrated approach will have to be pro-poor and centred on smallholders and the rural population, but with a concern to create opportunities for the newer generations and distribute the benefits of foreign investment. Adaptation policies are therefore at the forefront of any long-term response to strengthen the resilience of the population in Mozambique, in a context of climate emergency and particularly in extremely vulnerable territories such as Cabo Delgado.

References

Adger, W. N., & Kelly, P. M. (1999). Social vulnerability to climate change and the architecture of entitlements. *Mitigation and Adaptation Strategies for Global Change, 4*, 253–266.

Ahmed, K. (2021, March 6). 'So much trauma': Mozambique conflict sparks mental health crisis. *The Guardian.*

Amnesty International. (2021). *'What I Saw Is Death': War Crimes in Mozambique's Forgotten Cape.* Retrieved from London.

Araneda-Cabrera, R. J., Bermúdez, M., & Puertas, J. (2021). Assessment of the performance of drought indices for explaining crop yield variability at the national scale: Methodological framework and application to Mozambique. *Agricultural Water Management, 246*, 106692. https://doi.org/10.1016/j.agwat.2020.106692

Cunguara, B., & Hanlon, J. (2012). Whose wealth is it anyway? Mozambique's outstanding economic growth with worsening rural poverty. *Development and Change, 43*(3), 623–647. https://doi.org/DOI10.1111/j.1467-7660.2012.01779.x

FAO (2017). *The State of Food and Agriculture 2017: Leveraging food systems for inclusive rural transformation.* Food and Agriculture Organisation (FAO), Rome, Italy.

International Monetary Fund. (2014). *Poverty reduction strategy paper.* Maputo.

Matsinhe, D. M., & Valoi, E. (2019). *The genesis of insurgency in Northern Mozambique.*

Nussbaum, M. C. (2011). *Creating capabilities: The human development approach.* Belknap Press of Harvard University Press.

OCHA. (2020). *Mozambique Cyclone Kenneth: One Year After* [Press release]. https://reliefweb.int/report/mozambique/mozambique-cyclone-kenneth-25-april-2020-one-year-after

Otto, I. M., Reckien, D., Reyer, C. P. O., Marcus, R., Le Masson, V., Jones, L., Norton, A., & Serdeczny, O. (2017). Social vulnerability to climate change: a review of concepts and evidence. *Regional Environmental Change, 17*(6), 1651–1662. https://doi.org/10.1007/s10113-017-1105-9

Pelling, M., O'Brien, K., & Matyas, D. (2014). Adaptation and transformation. *Climatic Change, 133*(1), 113–127. https://doi.org/10.1007/s10584-014-1303-0

Salvucci, V., & Santos, R. (2020). Vulnerability to natural shocks: Assessing the Short-term impact on consumption and poverty of the 2015 flood in Mozambique. *Ecological Economics, 176,* 106713. https://doi.org/10.1016/j.ecolecon.2020.106713

Sen, A. (2013). The ends and means of sustainability. *Journal of Human Development and Capabilities, 14*(1), 6–20. https://doi.org/10.1080/19452829.2012.747492

UN News. (2020). *Mozambique Cabo Delgado violence is a 'desperate' situation, warns Bachelet* [Press release]. https://news.un.org/en/story/2020/11/1077612

USAID. (2019). *Food Assistance Fact Sheet: Mozambique.* https://www.usaid.gov/mozambique/food-assistance

Walker, B. H. (2020). Resilience: What it is and is not. *Ecology and Society, 25*(2). https://doi.org/10.5751/es-11647-250211

Wolff, J., & De-Shalit, A. (2007). *Disadvantage.* Oxford University Press.

World Bank. (2018). *Strong but not Broadly Shared Growth: Mozambique Poverty Assessment.* http://documents1.worldbank.org/curated/en/248561541165040969/pdf/Mozambique-Poverty-Assessment-Strong-But-Not-Broadly-Shared-Growth.pdf

Carla Gomes is a research fellow at the Institute of Social Sciences of the University of Lisbon, where she has collaborated in multiple projects and plans for climate adaptation and sustainability. Her current research interests are climate adaptation, human capabilities, environmental justice and local ecological knowledge.

Luísa Schmidt is a principal researcher at the Institute of Social Sciences of the University of Lisbon. She is coordinator of OBSERVA—Observatory of Environment, Territory and Society, and a member of the Scientific Committee of the Doctoral Program on Climate Change and Sustainable Development Policies (since 2009).

6

Climate Injustice in Brazil: What We Are Failing Towards a Just Transition in a Climate Emergency Scenario?

Pedro Henrique Campello Torres, Ana Lia Leonel, and Gabriel Pires de Araújo

Abstract This chapter questions how climate justice is—if it is—present in the Brazilian climate change debate and praxis. For this purpose, three climate change plans are analyzed concerning how categories related to justice, poverty and right issues are inserted or not inserted in current local planning practices. The chapter is illustrated by analysing three cities, Fortaleza, Rio de Janeiro and Santos. The Plans demonstrate progress regarding science and policy dialogue but still severe gaps concerning community participation. The authors warn that the absence of social participation, not only business as usual but also

P. H. Campello Torres (✉) · G. P. de Araújo
Institute of Energy and Environment, University of Sao Paulo, Sao Paulo, Brazil

G. P. de Araújo
e-mail: gabriel.pires.araujo@usp.br

A. L. Leonel
Federal University of ABC (UFABC), Santo André, Brazil

© The Author(s), under exclusive license to Springer Nature Switzerland AG 2021
P. H. Campello Torres and P. R. Jacobi (eds.), *Towards a just climate change resilience*, Palgrave Studies in Climate Resilient Societies, https://doi.org/10.1007/978-3-030-81622-3_6

breaking paradigms and including the most vulnerable, ends up rein-forcing unequal historical processes—the opposite direction towards equity and just a sustainable future.

Keywords Global South · Adaptation plans · Climate justice · Cities · Local level

Introduction

Climate justice is still an embryonic term in Brazil and Latin America (Torres et al., 2020). The notion, which emerged in the Global North, had its contours and essays from the South, with the Cochabamba Conference in Bolivia in 2010 (Turner, 2010). On the other hand, perhaps due to the further weakening of the support block of leftist leaders in Latin America—which some call the *pink tide* (Serrano, 2013)—the issue did not emerge as a priority political agenda in the region. The fact that the climate justice movement in Brazil is not solid or consolidated does not mean that the disproportionate impacts of climate change do not hit the most vulnerable populations more severely—those with the least capacity to react to the harm suffered. Kashwan argues that "Climate Action Movements" should not be under-stood to be synonymous with "Climate Justice Movements". (Kashwan, 2021), which makes perfect sense for the case of Latin America and probably for much of the Global South. Despite climate injustices—local or global—for the Brazilian case, the existence of two agendas is quite noticeable—one related to the "Climate Action Movements", here leveraged by international organizations and global networks, and the "Climate Justice Movements", originating from grassroots social groups, linked to the agenda of environmental justice.

This article reflects on this question: Where is climate justice in Brazil? To this end, it seeks to dialogue with other works that have already anal-ysed the non-incorporation of this agenda at the local level, whether in the National Climate Change Plan (Brazil, 2009) or the National Adap-tation Plan (Brazil, 2016). Instead of looking at the national level, the focus here will be the local level—the cities, a place always cited as the locus of concrete actions to mitigate and adapt to climate change.

The title already announces that we start from understanding what we are failing towards a just transition in a climate emergency scenario. The question, therefore, is to discuss the intricacies that justify this scenario. An analysis of what we have been doing concerning climate change assessments and policy responses for the Brazilian cities is necessary and urgent. This assessment is vital to prevent failure in planning for climate change impacts, excluding communities in the participatory process, or in not tackling inequalities. Extreme climate events—severe droughts, extreme rainfall, fires and floods—already strongly affect the Brazilian territory. The absence of just transition on planning adds barriers to implementing policies to prevent loss of life and increase the pattern of vulnerabilities and environmental inequalities in municipalities.

In that sense, this research seeks to contribute to the climate justice debate from the South, addressing climate change inequalities and their responses from an original and interdisciplinary perspective. Science and traditional knowledge already provide accurate data for policies to adapt to climate change impacts. However, decision-making remains a challenge, especially on a subnational scale (Torres et al., 2021). This article, therefore, aims to fill this gap by providing insights on subnational policies related to climate change in a scenario that more climate change plans should be elaborated by cities from all Brazilian regions in the near future. The fact that we have few plans, could be seen as an opportunity to design those new instruments towards a just sustainable transition.

In this sense, the present study contributes to addressing climate justice in Brazil by providing data and illustrative case examples that, on the one hand, could support decision-makers. On the other hand, it could inspire civil society, grassroots movements and traditional communities to strengthen this agenda. The article is divided as follows. After the introduction, we describe the materials and methods of our analysis. The next part of this article seeks to contextualize the climate justice debate and its approximations with the Brazilian case. Three cases on Brazilian cities and their climate change plans were chosen to subsidize arguments and debate empirically in the subsequence part. Then, a discussion based on the results found in the cases, in dialogue with the worked literature, is carried out to systematize the learnings and possibilities of pushing this agenda forward. In the final part of this chapter, we conclude and summarize the lessons learned towards the strength of this research agenda.

Material and Methods

To address the objectives of this chapter, the analysis of municipal climate change plans carried out in the last four years was realized. Despite its vulnerability to climate change and an imperative need for adaptation, Brazil has 5070 municipalities, and only thirteen have local climate change plans. Of these 13, we will work here with three cities: the plans of Fortaleza (Ceará state), Rio de Janeiro (Rio de Janeiro state) and Santos (São Paulo state). Figure 6.1 localizes these three cases in Brazil, all of them in the coast, one in the northeast and the other two in the southeast. Considering the vulnerability and risks, the map (Fig. 6.1) shows the areas of *Aglomerado Subnormal* (AS) (IBGE, 2010), which is a

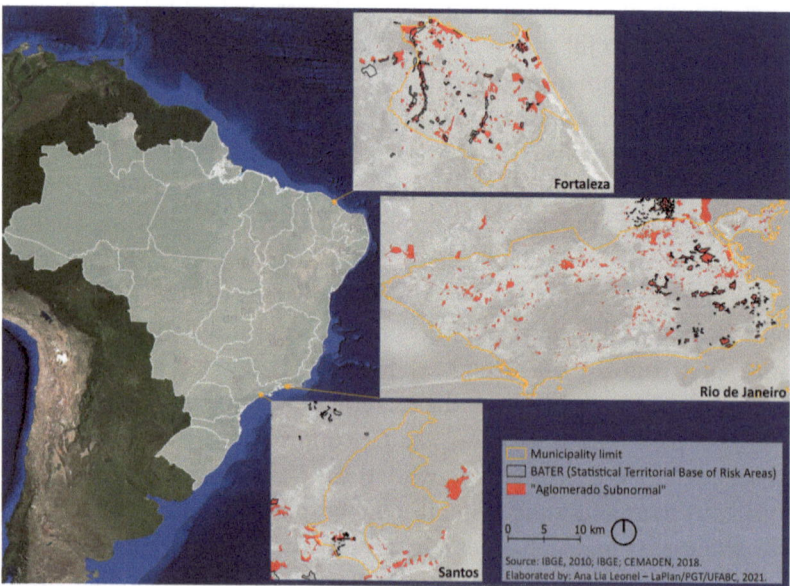

Fig. 6.1 Case Studies, Risk Areas and "Aglomerados Subnormais". Author's elaboration, 2021 (*Source* IBGE [2010] and IBGE; CEMADEN [2018])

territory typology for a form of irregular occupation of land (as *favelas*, or slums, for example),[1] and also the Statistical Territorial Base of Risk Areas (BATER, in Portuguese). The BATER associates territorialized sociodemographic data in Census sectors with the features of areas at risk of flooding, runoff and landslides (IBGE; CEMADEN, 2018).

In the case of the city of Rio de Janeiro, the instrument analysed was the Strategy for Adaptation to Climate Change in the City of Rio de Janeiro (SMAC / COPPE-UFRJ, December 2016). In Santos, the Municipal Climate Change Plan (PMMCS) was selected for analysis. And in Fortaleza, the construction of the Fortaleza Adaptation Plan in its development with the Climate Change Forum (FORCLIMA). An important clarification: The case of Fortaleza is not precisely a finished plan, as in the cases of Rio de Janeiro and Santos, but the construction of the plan. We believe that adding the case of a plan under construction helps us observe how the stages are being built, the forms and strategies of participation and other issues pertinent to our analysis.

The authors analysed the plans using specific criteria to verify how social issues related to climate change are being addressed by these local planning and governance instruments. Plan analysis is a possible procedure to operate methodologically among so many others, so assume that the procedure is exploratory and not definitive. Recognize, for example, the limitations for community and non-state actions that are not imbued with this type of frame. On the other hand, they are formal instruments in which the very arena for participation and knowledge of community actions/desires of non-state actors emerges.

Guiding questions for the analysis of the plans mentioned above:

(a) Historic. Is it the city's first climate plan? Which one first? When does it have a municipal climate policy? Does the city have a climate and adaptation plan/policy?

(b) Description. Are there traditional populations living in the municipality? If so, which ones and where? Are there any maps of that?

---

[1] Established by IBGE (Brazilian Institute of Geography and Statistics), https://www.ibge.gov.br/en/geosciences/territorial-organization/territorial-typologies/17553-subnormal-agglomerates.html?=&t=o-que-e, accessed on 22 April 2021.

What percentage of the population lives in subnormal agglomerations in the municipality (see IBGE).

(c) Multilevel. Are there climate/adaptation policies in the state? If so, does municipal policy dialogue with it? Climate Justice

(d) Do the descriptors "justice", "rights", "poverty", "inequality", "vulnerability", "vulnerable", "traditional populations" appear in the plan? If so, how many times? How these issues are addressed, whether they are in the plan.

(e) Climate changes. What are the main effects of climate change that these territories tends to feel? (Sea level rise, drought, more rain, landslides?).

(f) Plan responses. Brief analysis of the responses contained in the plan. They address a change in the status quo in city planning or look like traditional measures. Something that addresses traditional populations, slum dwellers, etc.? Something in general that draws attention?

(g) Participation. How was society's participation in the plan? From the start? With public consultation (if there was a consultation, observe how many changes occurred after the consultation) / public hearing (if there was an audience, how many and look at the presence of social groups)? Were sector meetings held with traditional populations or in areas of subnormal agglomerations (slums)?

Figure 6.2 graphically represents the methodological steps applied.

Theoretical Framework

In Brazil, the association between climate change and environmental tragedies is not apparent (Milanez; Fonseca, 2011), such as landslides, floods, droughts and fires. Much less than climate change and its consequent tragedies, it increases the population's vulnerability that is less able to deal with these consequences. Adaptation to climate change, risk, and disaster prevention must be closely linked to local development and decreases in inequality, especially in regions like Latin America. The

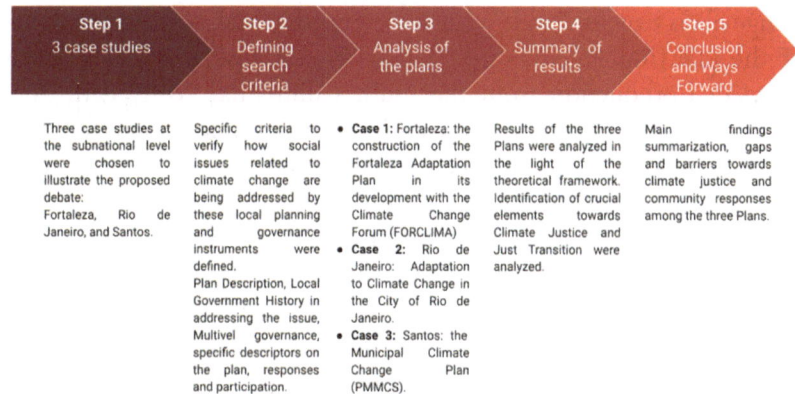

Step 1	Step 2	Step 3	Step 4	Step 5
3 case studies	Defining search criteria	Analysis of the plans	Summary of results	Conclusion and Ways Forward
Three case studies at the subnational level were chosen to illustrate the proposed debate: Fortaleza, Rio de Janeiro, and Santos.	Specific criteria to verify how social issues related to climate change are being addressed by these local planning and governance instruments were defined. Plan Description, Local Government History in addressing the issue, Multivel governance, specific descriptors on the plan, responses and participation.	• **Case 1:** Fortaleza: the construction of the Fortaleza Adaptation Plan in its development with the Climate Change Forum (FORCLIMA). • **Case 2:** Rio de Janeiro: Adaptation to Climate Change in the City of Rio de Janeiro. • Case 3: Santos: the Municipal Climate Change Plan (PMMCS).	Results of the three Plans were analyzed in the light of the theoretical framework. Identification of crucial elements towards Climate Justice and Just Transition were analyzed.	Main findings summarization, gaps and barriers towards climate justice and community responses among the three Plans.

Fig. 6.2 Methodological Steps (*Source* Author's elaboration, 2021)

identification of impacts, vulnerabilities, and, consequently, the assessment of adaptation actions must be part of the routine of the public manager, and civil society so that the produced knowledge presents clarity on any applicable measure based on technical-scientific scenarios.

But how to address this issue? Several authors have strained the dialogue between the problems related to environmental injustices (Anguelovski et al., 2019), the fight against inequalities and green privileges (Park & Pellow, 2011) and the importance of community engagement in elaborating answers to these questions (Agyeman et al., 2016; L. Shi et al., 2016). On the other hand, most of these studies, except for a few exceptions, are from the Global North. Or have as analysis of cases realities of that fraction of the planet.

In this sense, we seek here a circular and interdependent theoretical-methodological articulation, using questions of knowledge proper to the notion of environmental justice (Agyeman et al., 2016)—its articulation with climate justice, and questions from a Political Ecology approach—for whom, how, why, how, what? (Heynen et al., 2006; Meerow & Newell, 2016). These two fields of knowledge anchored in productions and practices from the South and the dialogue with public management/planning—see Fig. 6.2.

From the Political Ecology approach, it is imperative to consider the problems in terms of questioning as for "whom would those climate change plans benefit?" and "what is the purpose, method, time, and place for the plan implementation?" This approach needs to dialogue with knowledge production from the south (Alimonda, 2011; Leff, 2015; Souza, 2019) to contribute to the criticism exposed here about for whom and how this territory has been socially produced—according to the specific historical process of that region.

Swyngedouw (2009), in contrast to the environmental justice approach, argues that what is more vital than reflecting on the distribution of injustices is the debate on politics, power and the production of inequalities, because there is the arena where politics, power, privileges and inequity are structured and need to be disrupted for new egalitarian forms to emerge.

While equality is an "ontologically given condition of democracy" (Swyngedouw, 2009, p. 17), justice, on the other hand, "disappears from the terrain of the moral and enters the space of the political under the name of equality" (Swyngedouw, 2009, p. 15). However, our start point is that both tensions are necessary and that there is an essential dialogue between environmental justice and political ecology. That is, we consider that both strains are required and that there is a complementary dialogue between environmental justice and political ecology. A deeper questioning of power—and its unequal asymmetries—is fundamental to reduce inequalities and promote justice. As well, the lens that shows in the territory the distribution of inequalities and environmental privileges. Both are crucial tools for instrumentalizing social movements and providing subsidies for decision-making in the public sphere. This twofold movement, if carried out, contributes to the necessary radicalization of the democratic process. And to what the author understands as what should "emerge" in a "more egalitarian way".

In this chapter, we operate in an articulated way between these two fields of knowledge—Environmental Justice and Political Ecology (Fig. 6.3). We are questioning "who would those climate change plans benefit?" and "what is the purpose, method, time, and place for the plan implementation?" and, at the same time, we verify how formal planning—whose hold power—is dealing with these issues. Avoiding, perpetuating, or increasing these inequalities that are unevenly distributed across territories.

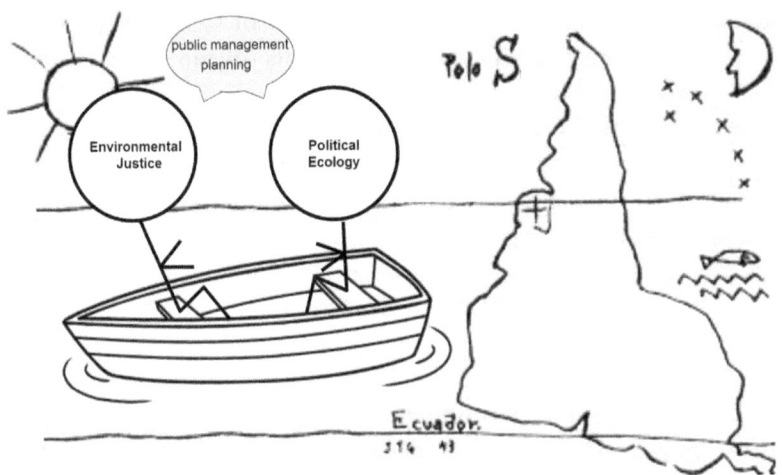

Fig. 6.3 Theoretical-methodological approach (*Source* Author's elaboration, 2021—Adapted from Joaquín Torres García "América Invertida")

It is precisely in the assessment and diagnosis of which communities are most exposed to the threats of climate change that the dialogue with environmental justice takes place in our proposed framework. So how do community groups, the most vulnerable groups, or traditional populations deal with this agenda in Brazil? Research has pointed out how the topic is still fragile on the agenda of social movements and communities. A possible explanation is that a priority agenda of demands for fundamental rights still has a long way to go (Torres et al., 2020).

Other factors that need to be better evidenced by empirical material would be the capturing of this agenda by big international non-governmental organizations, with neutralization of more local demands, which exposes contradictions inherent to the system in which these organizations are part of. A significant national example—and little-explored on Brazilian climate change literature—is the Carta de Belém from 2010 (Segebart & Konig, 2014). The manifest was signed by a hundred Brazilian organizations, including grassroots, social movements, unions, religious organizations and peasants, among others. In an unprecedented way, self-declared groups such as environmental organizations and movements, male and female workers in family and

peasant agriculture, agroextractivists, members of Quilombola communities, women's organizations, urban grassroots organizations, fishermen
and women, students, traditional peoples and communities and native
peoples sharing the struggle against deforestation and for environmental
justice in the Amazon and in Brazil at large. The Group—Grupo Carta
de Belém—positioned themselves incisively and clearly against market-
based mechanisms as tools to reduce carbon emissions based on the firm
conviction that the market cannot be expected to take responsibility for
life on the planet as the REDD mechanism.

The mention of social movements concerning the questions addressed
in this chapter is significant because both the approach for environmental
justice and its ramification, climate justice, operate in an integrated
manner with demands and denunciations of the most affected or most
exposed to environmental inequalities groups. In addition, in a climate
emergency scenario, social movements that defend the most vulnerable populations bring important contributions to face the problem
(Martinez-Alier et al., 2016; Rammê, 2012). This leads us to two questions: (1) community participation related to climate justice responses
towards equity and just sustainabilities paths. (2) The understanding that
climate justice is a ramification of the environmental justice approach.

The first makes us reflect again on the dialogue, which does not always
exist between political ecology and management/planning. As well as
the dialogue proposed for the integration of this approach with the lens
of environmental justice. In other words, are plans, policies and their
instruments being planned, executed and monitored, in order to reduce
environmental inequalities in the territory? If not, who is benefiting?
How and why? The second is aligned with what some authors have been
working on the transition of the classical environmental justice approach
(Bullard, 1990) with a local focus on dams and toxic waste, among other
dangers to which the most vulnerable communities are exposed, to a
focus centred on climate change (Agyeman et al., 2016; Porter et al.,
2020; Schlosberg et al., 2014).

If these two issues are not connected, the claim for climate justice will
be meaningless from a view from the South. This means understanding
the discrepancies mentioned above between demands for fundamental
rights—already minimally universalized in most of the Global North and

resolving issues related to structural racism as a result of a colonial past, a dominant elite based on serotinous oligarchic and retrograde capitalism. This means that these agendas must not operate and be strengthened from the bottom-up way. Either by large international non-governmental organizations or by state power.

Hurricane Katrina, which hit New Orleans in 2005, is seen as an important milestone for the movement's move from environmental justice to climate justice. But as Schlosber and Colins (2014) detail four years earlier, the Environmental Justice and Climate Change Initiative was founded in 2001 due to the first Climate Justice Summit at The Hague (Netherlands) during the UNFCCC/COP6.

> In 2005 Hurricane Katrina solidified the confluence of the environmental justice framework and the issue of climate change. Bullard and Wright start their important reflections on Katrina by laying out the preexisting injustices in the city of New Orleans—including the segregation, poverty, failing education system, and substandard housing. The community was underprepared, in the sense that infrastructure and living standards of minority populations were already vulnerable before the storm. And they were underserved after the storm in that they received less information, less government relief, fewer loans, and continued discrimination. (Schlosberg & Colins, 2014, p. 362)

The example mentioned above exposes, on the one hand, the intimate intersection between movements for environmental justice and climate justice. On the other hand, it explains how rationality from its founding landmark in the Global North, with most scholars and activists from countries and organizations in that region, may have been defining the climate justice trajectory agenda in the Global South.

Schlosberg and Collins (2014) demonstrate how climate justice movements are based on principles of social justice, democratic accountability, participation and ecological sustainability. What for the authors can also be seen in the environmental justice movements, "The two thriving grassroots movements have influenced each other, and even fused in many ways" (p. 370). The same cannot be said for the case of Brazil—or other territories in the Global South. Although we don't rule out that this may still occur—or even that transition is already occurring at this point. Future research will address this issue.

Illustrative Cases for Discussion

In this session, we discuss the three illustrative cases to verify how—if it is—the transition is occurring towards climate change adaptation in these places. This experimental verification is carried out based on the procedures indicated in the materials and methods section. In an articulated way with the theoretical framework proposed here, to understand who is being benefited—if any—by the researched plans and understand how the most vulnerable populations are being treated in this planning.

Case 1—City of Fortaleza Municipal Climate Change Plan and the Climate Change Forum (FORCLIMA)

The Adaptation Plan of the municipality of Fortaleza appears as in formulation according to information present on the city hall website. One of those responsible for formulating this plan is the Climate Change Forum (FORCLIMA). Established by Decree 13.639/2015, this forum is charged with instituting the Municipal Climate Change Plan and the Municipal Climate Change Policy, which includes adapting to climate change within the municipality. Among the activities already carried out by FORCLIMA, we highlight the contribution to the Low Carbon Urban Development Policy, sanctioned in 2017 and which should therefore dialogue with the subsequent local plans and climate policies.

The existence of international cooperation with the partnership between the Development Bank of Latin America and the French Development Agency in the project stands out in our analysis. This partnership takes place in the project "Ciudades y Cambio Climático", a project which encompasses several cities in Latin America, including the municipality of Fortaleza, where the Climate Change Vulnerability Index was developed in conjunction with local public management, contributing to the development of the Adaptation Plan.

Concerning the most vulnerable population exposed to harmful effects of climate change in the municipality of Fortaleza, the latest census reveals that 396,370 people (16.16% of the total population) live in a total of 194 AS (Fig. 6.1). A relevant part of the population that needs to be heard and prioritized by adaptation policies since they are the main affected.

It is noteworthy that the decree instituted FORCLIMA reinforces the importance of promoting and discussing effective actions to mitigate the effects of climate change at the federal, state and municipal levels. Besides, the decree also states the formulation of guidelines and norms of the Municipal Policy on Climate Change must be prepared aligned with the Interministerial Commission on Climate Change, the National Policy on Climate Change, the National Plan for Climate Change, the Brazilian Forum on Climate Change and the Ceará Forum on Climate Change—demonstrating the presence of a multilevel dialogue in the formulation of municipal climate policies.

The specific process of building the plan had five phases, which included: one-off consultations with specialists with on-site visits; Technical Table Formation; participatory workshops with the population; validation meetings, with technical table, population and representatives of the city hall present; one-off consultations with city hall specialists. Therefore, there is a governance perspective in the construction of the plan, although the stakeholder's present needs to be analysed in a more specific way to verify whether the meetings were representative concerning the population. Also noteworthy is that the population is currently not only in the survey and data collection stage for the construction of the plan but also in the validation process, making the process more participatory.

The construction of the Fortaleza Adaptation Plan identified the main climatic hazards for the municipality as extreme rains, temperature increase, rising sea level and drought. The plan's construction has been addressed among several strategies that seek to rely on social participation to address those climatic hazards. The ongoing social participation claims to be based on multiple actors and to value scientific knowledge and the popular knowledge of those who, according to the plan, are the ones who experience the effects of climate change. Participatory workshops were held with social actors to correlate climate change with the daily actions of the population. In the plan presentation from FORCLIMA, there is no qualification of these social actors (if they belong to traditional populations, residents of AS, etc.), which would be an essential issue when characterizing genuinely participatory process, and capable of responding to the wishes of the most vulnerable so that the plan does not amplify inequalities.

The plan also presents a territorial mapping of the climatic risks of Fortaleza, with an index of climatic risk to extreme rains, climatic risk index to temperature change, the climatic risk to rising sea levels, and climatic risk to droughts. These indexes were gathered and gave rise to the map of the current and future climate risk index of the municipality, identifying the current hotspots in 2040, determined by the vulnerability index to climate change, where the vulnerability aspect is dealt with more forcefully.

Case 2—The City of Rio de Janeiro Climate Change Adaptation Strategy

The city of Rio de Janeiro was one of the first cities in the country, after Palmas, São Paulo and Belo Horizonte, to establish a Municipal policy on Climate Change and Sustainable Development. The Policy set targets for reducing greenhouse gas emissions: 8% in 2012, 16% in 2016, 20% in 2020, based on the 2005 emissions recorded by the Greenhouse Gas Emissions Inventory.

Rio's CLIMATE CHANGE FORUM was created in 2009 (DECREE 31.415, of November 30, 2009) to monitor and construct the municipal climate policy. The Forum, in its description, establishes the composition by representative segments of the Public Power, the private initiative and civil society, whose objective is to contribute in the search for viable solutions for the adoption of public policies aiming to reduce and combat the climatic effects in the city.

In 2016 the Climate Change Adaptation Strategy was published during the *élan* of the *mega-events*—RIO + 20, World Cup, Olympics and Paralympics—that took place in the city of Rio de Janeiro and culminated in the current plunge into a financial, political and moral crisis (Cavalcanti et al., 2016; Sánchez & Broudehoux, 2013). The plan publication was part of the narrative that was sought to build by the City Hall of a green legacy, mainly from the Olympic Games of that year.

Climate change, therefore, had a central role among other factors: a coastal city, with 6.5 million inhabitants, high density in the lowlands and high rate of urbanization, with 1,035 slums, susceptibility to flooding, with fluvial flooding, islands and heatwaves, susceptibility to landslides of mass, average sea level, waves and oceanic agents, the vulnerability of beaches, lagoons and bays to waves and undertow as well as the mirror formed between urbanization and housing.

The plan was prepared based on the partnership between the Management of Climate Change and Susceptible Development of the Municipal Secretariat for the Environment (SMAC) and the Center for Integrated Studies on the Environment and Climate Change (Centro Clima/COPPE/UFRJ). In its presentation, the plan recognizes the participation of specialists as essential: technicians from the municipal and state secretariats and autarchies. There is no mention of social participation, traditional knowledge, or other community forms to develop a plan in a city where 22.15% live in subnormal agglomerations.

The plan is divided into sessions that allow us, in dialogue with the adopted framework, to question "Adapting is necessary," adapting for whom and how? "What awaits us," who are "us," are we the same in all areas of the city? "What threatens us"? Threat to everyone in the same way? "Our exposure and vulnerability," our who? And, finally, "Our way forward," which ways and where? The document states that it is a guide to provide subsidies for elaborating the Adaptation Plan for the city of Rio Janeiro. Which must and "established paths for adaptation that aim to ensure the protection of the built and natural heritage and preserve relations economic and socio-cultural factors in the face of climate change, for the benefit of current and future generations" (Janeiro, 2016, p. 9).

A crucial positive issue to be highlighted is the plan division structure into its existing planning areas, which allows a specific view of the territory and linked to a broader picture of the city. Also the results were specialized and already exposed by the municipality's Planning Areas (AP). In addition to bringing the theme closer to the current practice of planning and management, it also makes it possible for residents to know and recognize the specificities of these locals. At the same time, it recognizes something that should be current, but not right, that each fraction of the territory and its social fabric are heterogeneous and diverse.

The plan works on six strategic axes: (A) Strengthen institutional and human capacity; (B) Guarantee the conservation and integrity of ecosystems and the rational and sustainable use of natural resources; (C) Foster the promotion of the population's health in the face of climate change; (D) Conduct the occupation and use of the territory to promote urban-environmental quality; (E) Ensuring efficient urban mobility and sustainable; and (F) Guarantee the functioning of Strategic Infrastructures under adverse climatic conditions.

"Justice," "rights," "inequality," "traditional populations" are not mentioned in the plan. Poverty is mentioned once in the 90 pages related to extreme poverty and in a generic way in a paragraph about how cities, in general, are especially vulnerable to climate hazards—in a city where 16.41% are vulnerable to poverty and 1.25% extremely poor. On the other hand, "Vulnerability" associated with physical or social factors has 72 mentions, while the descriptor "Vulnerable" has only 1.

The plan recognizes that,

> Social inequity and insufficient housing policies are factors that, in the same way, aggravate vulnerability, by inducing the disadvantaged population to occupy areas at risk (slopes, areas prone to flooding), where, even, the provision of urban infrastructure and services is in deficit. On the other hand, it is essential to note that the higher income population also occupied risk areas, such as slopes, the marginal strip of lagoons, and coastal areas. The difference between both classes lies in the ability to deal with climatic dangers. (Janeiro, 2016, p. 14)

To address this issue, the plan proposes a vulnerability assessment integrating Urbanization and Housing, Urban Mobility, Health and Environmental Assets and what the plan calls strategic infrastructure. In the schedule presented in the plan, entitled "construction process," there is no mention of any participatory stage with society and communities in the cities.

The Plan section "Our way" is difficult to recognize which "our" we are dealing with concerning that inequality, fragmentation and the social production of space in Rio de Janeiro are historically marks of the urban evolution of the territory (Abreu, 2013). In this session, social and community participation is not mentioned in any of the strategic axes. Although in the "vision section" (p. 46), there is mention that it is necessary to increase resilience "so that the population has autonomy in their choices, understanding that the *Carioca* is the protagonist in building adaptive capacity to climate change" (Janeiro, 2016, p. 46).

Case 3—City of Santos and the Municipal Climate Change Plan (PMMCS)

The Municipal Climate Change Plan of Santos—PMMCS (in Portuguese) was published in 2016 (Decree nº 7.379), by efforts of the Municipal Commission for Adaptation to Climate Change—CMMC (in Portuguese), created by decree in November 2015 (Decree No. 7,293) and composed of representatives from several municipal departments, initially coordinated by the Municipal Secretariat for Urban Development and, from 2017, by the Secretariat for the Environment. The commission was charged with preparing the plan aligned with the National Plan for Adaptation to Climate Change, the State Policy on Climate Change and the Complementary Law No. 821/2013, which institutes the Master Plan for Urban Development and Expansion of the Municipality of Santos (our translation). The plan was the first one about adaptation in the country and, since 2018, Santos has participated in the pilot version of a Project to support Brazil in the Implementation of the National Agenda for Adaptation to Climate Change (ProAdapta)— a partnership between the Ministry of the Environment (MMA) and German Cooperation for Sustainable Development (GIZ). The PMMCS has also had the technical and scientific support of the Academic Advisory Committee since 2018. In 2018 and 2019, seminars were held to update the PMMCS with the participation of technicians from the city hall, researchers and specialists with studies on the coast of São Paulo (Carvalho, 2019; Santos, 2016; Santos' website).

Throughout the PMMCS, several passages reinforce the importance of public participation and clearly establish this as a principle. However, in the CMMC meeting minutes between the commission formation and the presentation of the PMMCS (in 2015 and 2016), it is clear that the population (more broadly) did not participate in the plan formulation. There were few guests registered in the minutes, and they are all scientists and researchers. According to news from the City Hall Portal and mentions from the minutes, at the beginning of 2020, workshops were held with the residents of Monte Serrat focused on adaptation actions. The Plan does not precisely propose to create programmes and projects that respond specifically to practical problems. This document intends to be a guide with objectives, goals and guidelines. Fulfilling this role, the Plan guides principles to be translated into adaptive actions and measures, emphasizing the vulnerable population and risk management based on intersectoral activities (Santos, 2016). In other words, the dynamics established is that the PMMCS was prepared by the CMMC with consultancy by researchers and scientists, in parallel with the construction of projects and practical action programmes for adaptation with the communities.

In the Plan, among the Thematic Axes (Section 8), specific to the Vulnerable Population (Section 8.4, p. 52), there is an understanding that, although climate change affects everyone, precarious communities and settlements (as *favelas*, for example), which already accumulate vulnerabilities, in addition to being the ones that least contribute to climate change, are the ones that suffer most from its consequences. This is because, in the case of Santos, these communities, in addition to having less access to services, are located in areas of high and very high geological risk and vulnerability (Santos, 2016, p. 53). In Santos, "residents of private households in the 24 AS represent 38,159 people (9.13% of the total population) in 10,767 households" (Santos, 2016, p. 56, our translation). Concerning Traditional Population, the PMMCS does not address any comment about it, but it does not mean that this is not an issue at the municipality. It is known that Santos has some traditional communities, for example the traditional *Caiçara* community of Diana Island. It is a lack of information, there is no Traditional Population Inventory in this region (Estado de São Paulo, 2019, p. 386).

The plan addresses and describes the various consequences of climate change in the municipality. Considering the vulnerabilities mentioned above, geological risks, such as landslides, are one of them. Besides, because it is a coastal municipality, effects related to the ocean are the most cited, such as rising sea levels, warming seawater temperatures and undertow—linked to strong waves, sea invasion, silting, erosion of the coast, saline intrusion. Also mentioned are heavy rains and, as a result, increased fog, the incidence of lightning, strong winds and flooding. The plan also addresses the health consequences of the population, such as the risks of contracting waterborne, food and vector-borne diseases and problems related to the temperature increase.

Analysing the PMMCS, we can say that there is some concern with climate and environmental justice, especially concerning the vulnerability of the poor population living in risk areas. Quantifying the descriptors selected for analysis in this article, we have justice, three occurrences; rights, five occurrences; poverty, two occurrences; inequality, two occurrences; vulnerability/vulnerable, 40 occurrences; traditional populations, zero occurrences.

In the plan, the term justice is related to the concept of climate justice, which argues that the consequences of climate change have a more significant impact on those who have contributed less and who are less able to deal with these effects. The descriptor vulnerability / vulnerable appears in three contexts: related to the environment, when the physical space itself is susceptible to climate changes, for example, "coastal zone vulnerability" (11 occurrences); related to the population, referring to those in precarious situations and poverty, considering that they are the ones that suffer most from the effects of climate change (7 occurrences); related to both, vulnerable areas and populations, referring at the same time to the vulnerabilities of both the physical environment and the population established in these spaces (22 occurrences).

Discussion

A first finding is that there is no up-to-date data available on traditional populations in Brazil, and still less their spatial distribution in the territory. Articulated with the proposed theoretical framework, we ask: For whom will this plan be? How will it benefit traditional populations exposed to the effects of climate change if there is no mapping of where they are, how many are there, what are their demands and cosmological understandings on climate change?

Table 6.1 helps us to understand the importance of social issues in the three cities studied. Fortaleza, for example, has 18.63% of its residents without access to water supply and almost 50% without sewage coverage. In the case of Rio de Janeiro, how not put the issue of AS in

Table 6.1 Assessments on cities specificities

Municipality	Fortaleza (%)	Rio de Janeiro (%)	Santos (%)
% urban population not connected to the water supply network (SNIS, 2017)	18.63	0.84	0
% of the urban population without sewage cover (SNIS, 2017)	49.28	14.02	0
% of the population living in informal settlement/favelas/AS (IBGE, 2010)	16.18	22.15	9.13
% of the residents declared as traditional communities	*	*	*
% vulnerable to poverty (IBGE, 2010)	32.88	16.41	8.08
% of extremely poor (IBGE, 2010)	3.36	1.25	0.60

*No data available
Note The data relating to the water supply network, population without sewage supply, vulnerability to poverty, and extremely poor were taken from the data aggregator Atlas do Brasil (www.atlasbrasil.org.br). The data referring to AS were taken from the IBGE aggregator (Cidades.ibge.gov.br)

which 22.15% of its inhabitants live as a strategic axis? More than that, how municipalities not listen to their demands, their needs, not only to validate the plan but to think about the concept and vision of what we intend to plan? The same for Santos, in which it presents the best socio-economic indicators among the three cities but still has about 10% of its residents living in AS and 8% vulnerable to poverty. In a planning process that seeks to articulate elements of Environmental Justice and Political Ecology, with management and planning practices, an effort in this chapter (Fig. 6.3), these issues must be at the forefront of any planning process aiming for climate justice with communities responses towards equity and just sustainability paths.

Fortaleza's adaptation plan is under formulation. That said, there is a possibility that aspects have not addressed yet, such as the presence of Climate Justice as a qualifier and driver of the Plan addressing vulnerabilities and social and environmental inequalities. Or characteristics that weaken the participatory nature of the Plan, such as non-identification of the social groups participating in the workshops, can still be inserted.

There are also strong points in the Plan, such as the dialogue between policies already formulated by other federal entities. The effort through participatory workshops to make the Plan participatory, and the participation of international agencies in elaborating the climate change index, which provided subsidies for the diagnosis made by FORCLIMA, should also be highlighted. These are actions that contribute to the recognition of the main effects negatives of the climate change process in the territory, at the same time that the most vulnerable populations are identified, crucial steps for the construction of an adaptation that faces the injustices related to the extreme events that tend to be exacerbated.

The analysed city plan of Rio de Janeiro presents several vital innovations. Certainly, it has the best scientific knowledge available from modelling, methods and instrumental tools. The effort for integrated local planning and strategic axes must be recognized. The critical link between policy and science is clear, which must have resulted in an essential experience for scholars and the practitioners involved. On the other hand, the articulation between science and policy shouldn't be detached from society. On the contrary, the academy should not legitimize initiatives in which the community has not been involved since the design concept. On the other hand, it should open the space, the necessary arena for community members to present their demands, indicate how they see the current situation and envision a future vision that is different from the current practice marked by profound inequalities.

Concerning Santos' case, the multilevel partnership to support the elaboration of climate change adaptation policies takes the results further. The ProAdapta, together with the Academic Advisory Committee, encourages and provides support for a complete diagnosis of the local reality and for thinking strategies and guidelines to face the challenges. Also as in Rio de Janeiro's case, the society's involvement at the elaboration phase of the plan is not clear. Analysing the minutes of the CMMC meetings, it seems that there was no participation of any social movements or organizations. On the other hand, the PMMCS is a guideline which foresees a second phase for the construction of projects and practical action programmes for adaptation with the communities. In this sense, this second phase has potential to go beyond the business-as-usual participation model and really involves the affected communities into the solutions.

Conclusion and Learnings

The present work contributes to reducing the knowledge gap between the climate justice agenda in Brazil, mainly in the possibility of articulation of the field of knowledge of Political Ecology, Environmental Justice in an applied science-policy dialogue. From the civil society point of view, the influence of robust and active participation of neither the "Climate Action Movements" nor the "Climate Justice Movements" are observed in the plans. The fact that this agenda is not consolidated opens a vast avenue for the occupation of this space by practices towards a just transition. On the other hand, avoiding that the theme would be captured only by green narratives that do not contribute to the resolution of historical inequalities.

An essential finding of the study is that data of traditional populations still a gap to be addressed. What hinders the elaboration of more plural plans and that aggregate different visions and diversities. As usual, social participation, especially outside the standard of business, remains a target to be achieved. In some cases, such as Rio de Janeiro, participatory practices are not even mentioned along with the plan.

On the other hand, in the three cases, an essential approximation between science-policy is perceived, seeking to use the best available science to elaborate scenarios, modelling, methodologies and indicators. It is aligned, in some way, with what Quay coined Anticipatory Governance to plan for climate change (Quay, 2010). The point is that Quay's formulation, when applied to the context of the Global South, will not find an appropriate anchor if it does not take into account community issues and the processes of unequal formation of societies (Travassos et al., 2020).

Table 6.2 summarizes comparatively the exploratory lessons learned from the illustrative cases. Such methodology can be replicated and improved by an emerging research agenda and serve both for future academic reflections and the practice of planning and management.

Table 6.2 Learning and findings from the cases

Case study	Fortaleza	Rio de Janeiro	Santos
Community involvement from the conception phase	Yes	No	Yes
Business-as-usual participatory process	Yes	NA	
Participatory process with communities	NA	No	
Recommendation of families relocation	NA	No	NA
Use of climatic modelling	Yes	Yes	Yes
Historical of climate change plans/policies	Yes	Yes	No
International cooperation	Yes	Yes	Yes
Multi-level articulation	Yes	Yes	Yes
Justice and Inequalities addressed	NA	No	Yes
Housing issues addressed	Yes	Yes	Yes
Climate Justice cited on the plan	NA	No	Yes

Source Author's elaboration, 2021

Acknowledgements The São Paulo Research Foundation (FAPESP) supported this work (2018/06685-9, 2019/18462-7 and 2015/03804-9).

References

Abreu, M. A. (2013). *A evolução urbana do Rio de Janeiro.* IPP.

Agyeman, J., Schlosberg, D., Craven, L., & Matthews, C. (2016). Trends and directions in environmental justice: From inequity to everyday life, community, and just sustainabilities. *Annual Review of Environment and Resources, 41,* 321–340. https://doi.org/10.1146/annurev-environ-110615-090052

Alimonda, H. (dir.). (2011). *La naturaleza colonizada: Ecología política y mineria en América Latina* (1a. ed.). Ciccus & CLACSO.

Anguelovski, I., Irazábal-Zurita, C., & Connolly, J. J. (2019). Grabbed urban landscapes: Socio-spatial tensions in green infrastructure planning in Medellín. *International Journal of Urban and Regional Research, 43,* 133–156. https://doi.org/10.1111/1468-2427.12725

Brazil. (2009, December 29). Federal Law No. 12,187. Establishes the National Policy on Climate Change (PNMC), and gives other Provisions. Presidency of the Republic. Civil House. Legal Sub-Office. http://www.planalto.gov.br/ccivil_03/_ato2007-2010/2009/lei/l12187.htm. Accessed on October 2021.

Brazil. (2016, May 10). Ministerial Ordinance MMA No. 150. Establishes the National Plan for Adaptation to Climate Change Climate, and gives other Provisions. http://www.mma.gov.br/images/arquivo/80182/Portaria%20PNA%20_150_10052016.pdf. Accessed on October 2021.

Bullard, R. D. (1990). *Dumping in Dixie: Race, class, and environmental quality.* Westview Press.

Cavalcanti, M., O'donnell, J., & de Sampaio, L. A. (2016). Futures and Ruins of an Olympic city. In B. Carvalho, M. Cavalcanti, & V. Rao (Org.), *Occupy all streets: Olympic urbanism and contested urban futures in Rio de Janeiro* (1 ed., Vol. 1, pp. 46–75). Terreform.

de Carvalho, D. A. (2019). Governança para a Conservação da Biodiversidade e Mudanças do Clima na Região Metropolitana da Baixada Santista. Dissertação (mestrado). Universidade Estadual Paulista (UNESP), Instituto de Biociências, São Vicente.

de Janeiro, R. (2016). Climate Change Adaptation Strategy for the City of Rio de Janeiro. Rio de Janeiro: Rio de Janeiro Prefeitura, Center for Integrated Studies on Climate Change and the Environment–Centro Clima/COPPE/UFRJ.

de Souza, M. L. (2019). *Ambientes e territórios: Uma introdução à Ecologia Política* (1. ed., Vol. 1, 350p). Bertrand Brasil.

Estado de São Paulo. (2019). *Plano de Manejo: Área de Proteção Ambiental Marinha Litoral Centro.* São Paulo. https://www.sigam.ambiente.sp.gov.br/sigam3/Default.aspx?idPagina=15388. Accessed on 20 April 2021.

Heynen, N., Kaika, M., & Swyngedouw, E. (2006). Urban political ecology. In N. Heynan, M. Kaika, & E. Swyngedouw (Eds.), *The nature of cities: Urban political ecology and the politics of urban metabolism.* Routledge.

IBGE. (2010). IBGE—Instituto Brasileiro de Geografia e Estatística. SIDRA—Sistema IBGE de recuperação automática. http://www.ibge.br/sidra/. Accessed on October 2021.

IBGE—Instituto Brasileiro de Geografia e Estatística; CEMADEN—Centro Nacional de Monitoramento e Alertas de Desastres Naturais. População em áreas de risco no Brasil. Rio de Janeiro: IBGE, 2018. https://biblioteca.ibge.gov.br/visualizacao/livros/liv101589.pdf. Accessed in April 2021.

Kashwan, P. (2021). Climate justice in the global North: An introduction. *Case Studies in the Environment.* https://doi.org/10.1525/cse.2021.1125003

Leff, E. (2015). Political ecology: A Latin American perspective. *Desenvolvimento e Meio Ambiente, 35,* 29–64.

Martinez-Alier, J., Temper, L., Del Bene, D., & Scheidel, A. (2016). Is there a global environmental justice movement? *The Journal of Peasant Studies, 43*(3), 731–755.

Meerow, S., & Newell, J. P. (2016) Urban resilience for whom, what, when, where, and why? *Urban Geography.*

Milanez, B., & Fonseca, I. F. F. (2011). *Justiça Climática e Eventos Extremos: Uma análise da percepção social no Brasil.* Revista Terceiro Mundo.

Park, L. S. H., & Pellow, D. (2011). *The Slums of Aspen: Immigrants vs the Environment in America's Eden.* New York University Press.

Porter, L., Rickards, L., Verlie, B., Bosomworth, K., Moloney, S., Lay, B., Latham, B., Anguelovski, I., & Pellow, D. (2020). Climate Justice in a Climate Changed World. *Planning Theory & Practice, 21*(2), 293–321. https://doi.org/10.1080/14649357.2020.1748959

Quay, R. (2010). Anticipatory governance—A tool for climate change adaptation. *Journal of the American Planning Association, 67*(4), 496–511.

Rammê, R. S. (2012). A política da justiça climática: Conjugando riscos, vulnerabilidades e injustiças decorrentes das mudanças climáticas. *Revista De Direito Ambiental, 17*(65), 367–389.

Sánchez, F., & Broudehoux, A. (2013). Mega-events and urban regeneration in Rio de Janeiro: Planning in a state of emergency. *International Journal of Urban Sustainable Development, 5*, 132–153.

Santos. (2015). *Decreto nº 7.293 de 30 de Novembro de 2015: Cria a Comissão Municipal de Adaptação à Mudança do Clima.* Diário Oficial de Santos: 01 de dez de.

Santos. (2016). *Estado da Arte do Plano Municipal de Mudanças do Clima de Santos (PMMCS).* Secretaria Municipal de Desenvolvimento Urbano. Comissão Municipal de Adaptação à Mudança do Clima.

Schlosberg, D., & Collins, L. B. (2014). From environmental to climate justice: Climate change and the discourse of environmental justice. *Wiley Interdisciplinary Reviews: Climate Change, 5*(3), 359–374. https://doi.org/10.1002/wcc.275

Segebart, D., & Konig, C (2014). Out of the forest—The climate movement in Brazil. In M. Dietz & H. Garrelts, *Routledge Handbook of the Climate Change Movement.* Routledge.

Serrano Hernández, R. (2013, October) *Giros culturales en la marea rosa de América Latina* (Tla-Melaua. Revista de Ciencias Sociales, [S.l.], n. 35).

Shi, L., Chu, E., Anguelovski, I., Aylett, A., Debats, J., Goh, K., Schenk, T., Seto, K. C., Dodman, D., Roberts, D., Roberts, J. T., & Van Deveer, S. D. (2016). Roadmap towards justice in urban climate adaptation research. *Nature Climate Change, 6*(2), 131–137.

Swyngedouw, E. (2009). The antinomies of the postpolitical city: In search of a democratic politics of environmental production. *International Journal of Urban and Regional Research, 33*, 601–620. https://doi.org/10.1111/j.1468-2427.2009.00859.x

Torres, P. H. C., Jacobi, P. R., Momm, S., & Leonel, A. L. (2021). Data and Knowledge matters—Urban Adaptation Planning in São Paulo (Brazil). *Urban Climate* (36). https://doi.org/10.1016/j.uclim.2021.100808

Torres, P. H. C., Leonel, A. L., Jacobi, P. R., & Araújo, G. (2020). Is the Brazilian national climate change adaptation plan addressing inequality? Climate and environmental justice in a global south perspective. *Environmental Justice, 13*(2), 42–46. https://doi.org/10.1089/env.2019.0043

Travassos, L., Torres, P. H. C., Giulio, G. M., Jacobi, P. R., Freitas, E. D., Siqueira, I., & Ambrizzi, T. (2020). Why do extreme events still kill in the

São Paulo Macrometropolis? Chronicle of a Death Foretold in the Global South. *International Journal of Urban Sustainable Development.*

Turner, T. E. (2010). From Cochabamba: A new internationale and manifesto for mother earth. *CNS, 21*, 56–74.

7

Community Practices and Climate Justice from the Global South: Synthesis and Ways Forward

Pedro Roberto Jacobi
and Pedro Henrique Campello Torres

Abstract This chapter presents the contributions present in each case study presented in the five different countries. It also gives ways forward, paths and possibilities for future research agenda to be explored.

Keywords Community practices · Paths to sustainability · Just transition · Global South · Climate justice

The chapters presented in this book represent an original approach to a better understanding of the community practices and climate justice in all continents within a Global South perspective. The five cases represent specificities but also convergencies related to resilience, adaptation and the importance of active participation of communities. Climate change is in all cases a driver and in each of the examples the emphasis

P. R. Jacobi (✉) · P. H. Campello Torres
Institute of Energy and Environment, University of Sao Paulo, Sao Paulo, Brazil

P. H. Campello Torres and P. R. Jacobi (eds.), *Towards a just climate change resilience*, Palgrave Studies in Climate Resilient Societies,
https://doi.org/10.1007/978-3-030-81622-3_7

is on the dynamics of organization to reduce its impacts and promote alternatives that involve communitarian initiatives that represent possibilities to hinder the different realities pictured in Bangladesh, Brazil, Mozambique, Solomon Islands and Uruguay.

The diagnosis varies from case to case, in **Bangladesh**, where increased incidences of different climatic shocks and stresses have a significant impact on the availability of potable water from both surface and groundwater sources in coastal areas, with direct reach on water security. Gradual increase in temperature due to climate change, together with changes in rainfall pattern lead to scarcity of freshwater during the dry season, thus provoking degradation of water quality with increased salinization and arsenic. To these natural factors, there has to be added a context of uncoordinated policy making and service delivery, lack of accountability and poor management of water resources and infrastructure.

In **Brazil**, the authors situate the difficulties that the Climate Justice agenda has encountered in the country. Through the analysis of plans, policies and instruments, the chapter is illustrated by analysing the case of three cities, Fortaleza, Rio de Janeiro and Santos. The Plans demonstrate progress regarding science and policy dialogue but still severe gaps concerning community participation. The authors warn that the absence of social participation, not only business as usual but breaking paradigms and including the most vulnerable, ends up reinforcing unequal historical processes—the opposite direction towards equity and just sustainable future.

The case in **Moçambique** is centred in Cabo Delgado in its Northern region, rich in environmental and mineral resources, that faces a combination of risks in the last few years, characterized by rising climatic and political instability, in a region already affected by persistent poverty. The region was strongly affected by recurrent tropical storms since 2019 with direct impact on the most vulnerable and deprived of human capabilities. The situation is worsened by the rising of political violence due to the expansion of Islamist group Al-Shabaab since October 2017, whose breeding is a direct reaction to the unequal distribution of benefits from foreign investments and the lack of opportunities for the younger generations, in a context of persistent poverty.

Solomon Islands, an archipelago comprising over 900 islands, face an already known situation as Pacific Island nations are on the front lines of some of the worst climate change impacts. Sea level and temperature rise threaten settlements, subsistence agriculture and fishing and exacerbate already vulnerable forest, marine and biodiverse ecosystems. Logging with effects on deforestation due to illegal activities has become a central issue of concern, characterized by denunciation of systemic fraud, and corruption of the largest logging companies. This has generated selective incentives causing on-going intra-tribal tensions and outcomes that divide communities and drive adverse social and climate change impacts that threaten the culture and traditions of indigenous populations and a rise in gender-based violence caused by disruption in traditional livelihoods and cultural practises.

In **Uruguay,** climate change is seen by small-scale fishers of different coastal locations, one of the drivers that can intensify the current fisheries crisis of declining catches as more frequent wind storms or unfavourable onshore wind conditions, which have led to a reduction in the number of fishable days. Inn Piriapolis, a coastal tourist city located, less than 100 km from the capital of the country, Uruguay), small-scale fishing has always been an important economic activity throughout the year. The number of small-scale fishers has decreased over the years, and nowadays, the activity is not more economically profitable, and those who remain in the activity have managed to diversify their fish sales methods as well as their sources of income.

These cases reveal notwithstanding its specificities, the impact of climate change in local and regional scale, but they also present different community practices and bottom-up processes that can be characterized as a combination of initiatives that contribute to the development of their adaptive capacity in an articulate and innovative process. Within the different realities, some aspects converge as the multiple stakeholders organizations and participatory dynamics as PAR in Uruguay and PRA in Bangladesh where these practices are based on social learning to stimulate capacity to generate and process new information about and their possible solutions. In the Solomon Islands emphasizing an environmental justice framework to assess some of the policy and planning responses to the challenge of climate change, same approach elaborated for the Brazilian case. In the case of Moçambique, the need of an integrated approach is strained by the authors as a step to move forward.

Within these different processes, the outcomes observed represent a range that varies according to the organizational capacity and replication opportunities. In Bangladesh as an outcome of the PRA, the community people suggested their own way towards resilience in terms of water security, highlighting a set of initiatives that combine articulation of NGOs and local authorities to lessen the effects of water insecurity. This implies the need to look deeper into the current absorptive and adaptive capacity of vulnerable communities and the different indicators that range from institutional setting to socio-environmental vulnerabilities to be addressed to promote real transformations in people's lives. Solomon Islands represents an interesting case of the role of environmental justice frameworks to identify the significance of capabilities and well-being in shaping local-level experiences emphasizing climate change responses that might foster just and resilient territories, as an example of best practise for the implementation of carbon market projects in a small state. The engagement of government with international organizations, and at the local level, socially and culturally invested NGOs is providing a positive strategy for carbon projects, focusing on several issues such as education, forest protection and alternative livelihoods in anticipation to uncertain returns from carbon markets.

In Uruguay, the creation of a multi-stakeholder group through PAR involves a range of social actors as small-scale fishers, scientists, government and non-government, with the purpose of addressing local problems of the fishery, within a transdisciplinary approach contributed to the development of adaptive capacity and climate resilience. Several critical domains were positively impacted, the learning process was approved, trust improved and bottom-up institutional changes were triggered. It strengthens a process based on fostering learning and co-production of knowledge, reducing sense of uncertainty and promoting the resilience of social-ecological systems. The empowerment of fishers involves multiple complexities and the challenge of engagement of the different actors for bridging the adaptation gap.

Moçambique represents the most institutional fragile case as the threat of political instability and the convergence of factors of socio-environmental, and most notably the climate emergency, requires an integrated socio-environmental approach, towards a transformative adaptation that fully addresses the root causes of vulnerability. The challenge at the national level is that the current national plans for adaptation (NAPs), demand cooperation within a perspective of development of those institutions, NGOs, initiatives and funding programmes, to reduce the regional disparities, social inequities and environmental injustice. In this direction adaptation, policies are seen as the forefront of any long-term response to strengthen the resilience of the population in a context of climate emergency, and this is relevant in the extremely vulnerable territory of Cabo Delgado.

Strengthening the role of communities and multi-stakeholder coalitions is represented within the articles as an essential articulation to reduce and in the most optimistic perspective the injustices and inequalities inherent to anthropogenic climate change, due to omission facing deforestation, contamination of water, declining catches and other economic impacts. These cases illustrate the challenge of countries in the Global South to cope with climate injustice and reduce risks of aggravation of existing vulnerabilities and promote alternatives that generate social inclusion, transparency of policies adopted and reduction of risks in its different scales and reaches.

We hope that this book can inspire future research, new books, new debates and new practices for planners for climate action and managers and decision-makers in the face of the climate emergency we are experiencing.

Index

© The Editor(s) (if applicable) and The Author(s), under exclusive
license to Springer Nature Switzerland AG 2021
P. H. Campello Torres and P. R. Jacobi (eds.), *Towards a just climate
change resilience*, Palgrave Studies in Climate Resilient Societies,
https://doi.org/10.1007/978-3-030-81622-3